Information and Instru

This shop manual contains several sections each covering a specific group of wheel type tractors. The Tab Index on the preceding page can be used to locate the section pertaining to each group of tractors. Each section contains the necessary specifications and the brief but terse procedural data needed by a mechanic when repairing a tractor on which he has had no previous actual experience.

Within each section, the material is arranged in a systematic order beginning with an index which is followed immediately by a Table of Condensed Service Specifications. These specifications include dimensions, fits, clearances and timing instructions. Next in order of arrangement is the procedures paragraphs.

In the procedures paragraphs, the order of presentation starts with the front axle system and steering and proceeding toward the rear axle. The last paragraphs are devoted to the power take-off and power lift systems. Intersper... where ...ed are addi... tabular specifications p... ...ning to w... ...ing, etc.

HOW TO USE THE INDEX

Suppose you want to know the procedure for R&R (remove and reinstall) of the engine camshaft. Your first step is to look in the index under the main heading of ENGINE until you find the entry "Camshaft." Now read to the right where under the column covering the tractor you are repairing, you will find a number which indicates the beginning paragraph pertaining to the camshaft. To locate this wanted paragraph in the manual, turn the pages until the running index appearing on the top outside corner of each page contains the number you are seeking. In this paragraph you will find the information concerning the removal of the camshaft.

More information available at haynes.com
Phone: 805-498-6703

Haynes Group Limited
Haynes North America, Inc.

ISBN-10: 0-87288-433-3
ISBN-13: 978-0-87288-433-5

Cover art by Sean Keenan

Disclaimer

SHOP MANUAL
FORD

Models 1100-1110-1200-1210-1300-1310
1500-1510-1700-1710-1900-1910-2110

The tractor model number, serial number and engine number are stamped on an identification tag located on left side of transmission housing.

INDEX (By starting Paragraph)

INDEX (CONT.)

DUAL DIMENSIONS

This service manual provides specifications in both the Metric (SI) and U.S. Customary systems of measurement. The first specification is given in the measuring system used during manufacture, while the second specification (given in parenthesis) is the converted measurement. For instance, a specification of "0.28 mm (0.011 inch)" would indicate that the equipment was manufactured using the metric system of measurement and the U.S. equivalent of 0.28 mm is 0.011 inch.

CONDENSED SERVICE DATA

	1100	1110	1200	1210
GENERAL				
Engine Make		Shibura		
Engine Model	LEK752C2	LEK757C	LEK802D	S723
Number of Cylinders	2	2	2	3
Bore	75 mm (2.95 in.)	75 mm (2.95 in.)	80 mm (3.15 in.)	72 mm (2.83 in.)
Stroke	80 mm (3.15 in.)	80 mm (3.15 in.)	80 mm (3.15 in.)	72 mm (2.83 in.)
Displacement	706 cc (43.1 cu. in.)	706 cc (43.1 cu. in.)	804 cc (49.1 cu. in.)	879 cc (53.6 cu. in.)
Compression Ratio	23:1	23:1	23:1	24:1
TUNE-UP				
Firing Order	2-1	2-1	2-1	1-2-3
Valve Clearance – Cold				
Intake	0.30 mm (0.012 in.)	0.20 mm (0.008 in.)	0.30 mm (0.012 in.)	0.20 mm (0.008 in.)
Exhaust	0.30 mm (0.012 in.)	0.20 mm (0.008 in.)	0.30 mm (0.012 in.)	0.20 mm (0.008 in.)
Valve Face Angle		45°		
Valve Seat Angle		45°		
Injection Timing, Static-BTDC	23°-24°	20°-21°	23°-24°	24°
Injector Opening Pressure	11760 kPa (1705 psi)	11760 kPa (1705 psi)	11760 kPa (1705 psi)	11760 kPa (1705 psi)

CONDENSED SERVICE DATA (CONT.)

TUNE-UP (CONT.)	1100	1110	1200	1210
Governed Speeds – Engine Rpm				
Low Idle	750-850	750-850	750-850	750-850
High Idle (No Load)	2750-2800	2750-2800	2850-2900	2850-2900
Rated (Full Load)	2600	2600	2700	2700
Power Rating at Pto				
Shaft	8.2 kW (11 hp)	8.6 kW (11.5 hp)	10 kW (13.5 hp)	10 kW 13.5 hp)
Battery				
Volts	12			
Ground Polarity	Negative			

CAPACITIES				
Cooling System	3.0 L (3.2 U.S. qt.)	2.5 L (2.6 U.S. qt.)	4.0 L (4.2 U.S. qt.)	2.3 L (2.5 U.S. qt.)
Crankcase*	3.3 L (3.5 U.S. qt.)	3.3 L (3.5 U.S. qt.)	4.0 L (4.2 U.S. qt.)	3.3 L (3.5 U.S. qt.)
Fuel Tank	14 L (3.7 U.S. gal.)	18 L (4.8 U.S. gal.)	14 L (3.7 U.S. gal.)	18 L (4.8 U.S. gal)
Standard Transmission & Rear Axle	18.9 L (20 U.S. qt.)	17 L (18 U.S. qt.)	18.9 L (20 U.S. qt.)	17 L (18 U.S. qt.)
Hydrostatsic Transmission & Rear Axle	....	15.5 L (16.4 U.S. qt.)	....	15.5 L (16.4 U.S. qt.)
Front Axle Differential Case	1.5 L (1.6 U.S. qt.)	1.5 L (1.6 U.S. qt.)	1.5 L (1.6 U.S. qt.)	1.5 L (1.6 U.S. qt.)
Front Axle Reduction Case (Each)	0.2 L (0.21 U.S. qt.)	0.2 L (0.21 U.S. qt.)	0.2 L (0.21 U.S. qt.)	0.2 L (0.21 U.S. qt.)

*With filter change.

SPECIAL TORQUES				
Connecting Rod Caps	24-27 N·m (18-20 ft.-lbs.)	24-27 N·m (18-20 ft.-lbs.)	24-27 N·m (18-20 ft.-lbs.)	29-34 N·m (22-25 ft.-lbs.)
Main Bearing Holders	....	....	71-81 N·m (52-60 ft.-lbs.)	25-29 N·m (18-25 ft.-lbs.)
Crankshaft Rear Plate	46-54 N·m (34-40 ft.-lbs.)	46-54 N·m (34-40 ft.-lbs.)	....	....
Flywheel	343-441 N·m (253-325 ft.-lbs.)	343-441 N·m (253-325 ft.-lbs.)	343-441 N·m (253-325 ft.-lbs.)	56-69 N·m (43-51 ft.-lbs.)
Cylinder Head	146-152 N·m (108-112 ft.-lbs.)	128 N·m (94 ft.-lbs.)	150-155 N·m (110-114 ft.-lbs.)	48 N·m (35 ft.-lbs.)

CONDENSED SERVICE DATA

	1300	1310	1500	1510
GENERAL				
Engine Make		Shibura		
Engine Model	LEK802D	S753	LET862C	K773
Number of Cylinders	2	3	2	3
Bore	80 mm	75 mm	85 mm	77 mm
	(3.15 in.)	(2.95 in.)	(3.35 in.)	(3.03 in.)
Stroke	80 mm	72 mm	100 mm	80 mm
	(3.15 in.)	(2.83 in.)	(3.94 in.)	(3.15 in.)
Displacement.	804 cc	954 cc	1134 cc	1117 cc
	(49.1 cu. in.)	(58.2 cu. in.)	(69.2 cu. in.)	(68.2 cu. in.)
Compression Ratio	23:1	23:1	21:1	23:1
TUNE-UP				
Firing Order	2-1	1-2-3	2-1	1-2-3
Valve Clearance-Cold				
Intake.	0.30 mm	0.20 mm	0.30 mm	0.20 mm
	(0.012 in.)	(0.008 in.)	(0.012 in.)	(0.008 in.)
Exhaust	0.30 mm	0.20 mm	0.30 mm	0.20 mm
	(0.012 in.)	(0.008 in.)	(0.012 in.)	(0.008 in.)
Valve Face Angle		45°		
Valve Seat Angle		45°		
Injection Timing,				
Static-BTDC	23°-24°	20°-21°	23°-24°	22°
Injector Opening Pressure	11760 kPa	11760 kPa	11760 kPa	11760 kPa
	(1705 psi)	(1705 psi)	(1705 psi)	(1705 psi)
Governed Speeds-Engine Rpm				
Low Idle	750-850	750-850	750-850	750-850
High Idle (No Load)	2900-2950	2950-3000	2650-2700	3000-3050
Rated (Full Load).	2700	2800	2500	2800
Power Rating at Pto				
Shaft.	10 kW	12.3 kW	12.7 kW	14.7 kW
	(13.5 hp)	(16.5 hp)	(17 hp)	(19.7 hp)
Battery				
Volts		12		
Ground Polarity		Negative		
CAPACITIES				
Cooling System	4.0 L	2.7 L	5.3 L	3.0 L
	(4.2 U.S. qt.)	(2.8 U.S. qt.)	(5.6 U.S. qt.)	(3.2 U.S. qt.)
Crankcase*	4.3 L	3.8 L	4.3 L	4.0 L
	(4.5 U.S. qt.)	(4.0 U.S. qt.)	(4.5 U.S. qt.)	(4.2 U.S. qt.)
Fuel Tank	22 L	26.6 L	22 L	26.6 L
	(5.8 U.S. gal.)	(7 U.S. gal.)	(5.8 U.S. gal.)	(7 U.S. gal.)
Transmission, Rear Axle &				
Hydraulic System	20 L	18 L	20 L	18 L
	(21 U.S. qt.)	(19 U.S. qt.)	(21 U.S. qt.)	(19 U.S. qt.)
Front Axle Differential Case	1.5 L	2.4 L	2.4 L	2.4 L
	(1.6 U.S. qt.)	(2.5 U.S. qt.)	(2.5 U.S. qt.)	(2.5 U.S. qt.)
Front Axle Reduction				
Case (Each)	0.18 L	0.22 L	0.22 L	0.22 L
	(0.19 U.S. qt.)	(0.23 U.S. qt.)	(0.23 U.S. qt.)	(0.23 U.S. qt.)
*With filter change.				
SPECIAL TORQUES				
Connecting Rod Caps	25-28 N·m	30-34 N·m	80-85 N·m	25-27 N·m
	(18-20 ft.-lbs.)	(22-25 ft.-lbs.)	(59-63 ft.-lbs.)	(18-20 ft.-lbs.)
Main Bearing Holders	....	25-29 N·m	....	48-53 N·m
		(18-22 ft.-lbs.)		(36-39 ft.-lbs.)
Crankshaft Rear Plate	46-54 N·m	27-33 N·m	46-54 N·m	46-54 N·m
	(34-40 ft.-lbs.)	(20-24 ft.-lbs.)	(34-40 ft.-lbs.)	(34-40 ft.-lbs.)
Flywheel	343-441 N·m	59-69 N·m	343-441 N·m	59-69 N·m
	(253-325 ft.-lbs.)	(44-50 ft.-lbs.)	(253-325 ft.-lbs.)	(44-50 ft.-lbs.)

CONDENSED SERVICE DATA (CONT.)

SPECIAL TORQUES (CONT.)	1300	1310	1500	1510
Cylinder Head	150-155 N·m (110-114 ft.-lbs.)	48 N·m (35 ft.-lbs.)	150-155 N·m (110-114 ft.-lbs.)	†

†61 N·m (45 ft.-lbs.) with 10 mm bolts; 95 N·m (70 ft.-lbs.) with 12 mm bolts.

CONDENSED SERVICE DATA

	1700	1710	1900	1910	2110
GENERAL					
Engine Make			Shibura		
Engine Model	LE892	H843	LEM853	T853A	T854B
Number of Cylinders	2	3	3	3	4
Bore	90 mm (3.54 in.)	84 mm (3.31 in.)	85 mm (3.35 in.)	85 mm (3.35 in.)	85 mm (3.35 in.)
Stroke	100 mm (3.94 in.)	84 mm (3.31 in.)	84 mm (3.31 in.)	100mm (3.94 in.)	100mm (3.94 in.)
Displacement	1272 cc (77.7 cu. in.)	1396 cc (85.2 cu. in.)	1429 cc (87.2 cu. in.)	1702 cc (103.8 cu. in.)	2268 cc (138.4 cu. in.)
Compression Ratio	21:1	23:1	21:1	21:1	21:1
TUNE-UP					
Firing Order	2-1	1-2-3	1-2-3	1-2-3	1-3-4-2
Valve Clearance-Cold					
Intake	0.30 mm (0.012 in.)	0.20 mm (0.008 in.)	0.30 mm (0.012 in.)	0.30 mm (0.012 in.)	0.30 mm (0.012 in.)
Exhaust	0.30 mm (0.012 in.)	0.20 mm (0.008 in.)	0.30 mm (0.012 in.)	0.30 mm (0.012 in.)	0.30 mm (0.012 in.)
Valve Face Angle			45°		
Valve Seat Angle			45°		
Injection Timing, Static-BTDC	20°-22°	22½°-23½°	26°-27°	23½°-24½°	23½°-24½°
Injector Opening Pressure			11760 kPa (1705 psi)		
Governed Speeds-Engine Rpm					
Low Idle	750-850	750-850	750-850	750-850	750-850
High Idle (No Load)	2600-2650	2825-2875	2900-2950	2650-2700	2650-2700
Rated (Full Load)	2500	2700	2800	2500	2500
Power Rating at Pto					
Shaft	17.4 kW (23.3 hp)	17.8 kW (23.9 hp)	20 kW (26.9 hp)	21.3 kW (28.6 hp)	25.9 kW (34.8 hp)
Battery					
Volts			12		
Ground Polarity			Negative		
CAPACITIES					
Cooling System	5.3 L (5.6 U.S. qt.)	5.5 L (5.8 U.S. qt.)	6.8 L (7.2 U.S. qt.)	7.0 L (7.4 U.S. qt.)	8.5 L (9.1 U.S. qt.)
Crankcase*	5.0 L** (5.3 U.S. qt.)	5.3 L (5.6 U.S. qt.)	5.5 L (5.8 U.S. qt.)	6.5 L (6.9 U.S. qt.)	7.5 L (7.9 U.S. qt.)
Fuel Tank	22 L (5.8 U.S. gal.)	29 L (7.6 U.S. gal.)	29 L (7.6 U.S. gal.)	35 L (9.3 U.S. gal.)	40 L (10.6 U.S. gal.)

CONDENSED SERVICE DATA (CONT.)

CAPACITIES (CONT.)	1700	1710	1900	1910	2110
Transmission, Rear Axle & Hydraulic System	22 L (23.2 U.S. qt.)	18 L (19 U.S. qt.)	24 L (25.4 U.S. qt.)	28 L (29.6 U.S. qt.)	32.2 L (34 U.S. qt.)
Rear Axle Final Drive Case (Each)	....	2.4 L (2.5 U.S. qt.)	....	....	....
Front Axle Differential Case	2.4 L (2.5 U.S. qt.)	3.3 L (3.5 U.S. qt.)	2.4 L (2.5 U.S. qt.)	4.2 L (4.5 U.S. qt.)	5.2 L (5.5 U.S. qt.)
Front Axle Reduction Case (Each)	0.22 L (0.23 U.S. qt.)	0.22 L (0.23 U.S. qt.)	0.22 L (0.23 U.S. qt.)	0.22 L (0.23 U.S. qt.)	0.22 L (0.23 U.S. qt.)

* With filter change.
** Crankcase capacity is 0.5 L (0.53 U.S. quarts) less when equipped with front wheel drive.

SPECIAL TORQUES					
Connecting Rod Caps	80-85 N·m (59-63 ft.-lbs.)	45-50 N·m (32-36 ft.-lbs.)	45-50 N·m (32-36 ft.-lbs.)	78-83 N·m (58-62 ft.-lbs.)	78-83 N·m (58-62 ft.-lbs.)
Main Bearing Holders	....	48-53 N·m (36-39 ft.-lbs.)	71-81 N·m (52-60 ft.-lbs.)	71-81 N·m (52-60 ft.-lbs.)	71-81 N·m (52-60 ft.-lbs.)
Crankshaft Rear Plate	46-54 N·m 34-40 ft.-lbs.)	46-54 N·m (34-40 ft.-lbs.)	46-54 N·m (34-40 ft.-lbs.)	46-54 N·m (34-40 ft.-lbs.)	46-54 N·m (34-40 ft.-lbs.)
Crankshaft Pulley	49-59 N·m (36-43 ft.-lbs.)	49-59 N·m (36-43 ft.-lbs.)	49-59 N·m (36-43 ft.-lbs.)	49-59 N·m (36-43 ft.-lbs.)	49-59 N·m (36-43 ft.-lbs.)
Flywheel	343-441 N·m (253-325 ft.-lbs.)	343-441 N·m (253-325 ft.-lbs.)	343-441 N·m (253-325 ft.-lbs.)	343-441 N·m (253-325 ft.-lbs.)	343-441 N·m (253-325 ft.-lbs.)
Cylinder Head	150-155 N·m (110-114 ft.-lbs.)	Note 1	Note 2	95 N·m (70 ft.-lbs.)	95 N·m (70 ft.-lbs.)

Note 1: 61 N·m (45 ft.-lbs.) with 10 mm bolts; 129 N·m (95 ft.-lbs.) with 14 mm bolts.
Note 2: 150-155 N·m (110-114 ft.-lbs.) for 11 large nuts and 58-62 N·m (43-46 ft.-lbs.) for 6 small nuts.

FRONT AXLE AND STEERING SYSTEM

FRONT AXLE (TWO WHEEL DRIVE)

All Models So Equipped

1. The front axle may be fixed tread width type or adjustable type for 1100, 1110, 1200, 1210, 1300, 1310, 1500 and 1510 models as shown in Figs. 1, 2 and 3. The adjustable axle used on 1700, 1710, 1900 and 1910 models is shown in Fig. 4. Adjustable axle used on 1710 Offset tractor is shown in Fig. 5, and adjustable axle used on 2110 tractor is shown in Fig. 6.

Front wheel toe-in is set by adjusting the length of the tie rod. Toe-in should be 0-5 mm (0-3/16 inch) on all models.

Clearance between axle pivot shaft and bushings (26 – Figs. 1, 2, 3, 4, 5 and 6) should be 0.02-0.15 mm (0.001-0.006 inch). Bushings should be renewed if clearance exceeds 0.30 mm (0.012 inch).

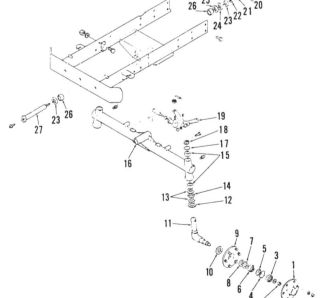

Fig. 1 – Exploded view of fixed tread front axle assembly used on two wheel drive 1100, 1110, 1200 and 1210 models.

1. Front wheel hub outer half
2. Nut
3. Outer bearing
4. "O" ring
5. Collar
6. Inner bearing
7. Seal
8. Seal
9. Wheel hub inner half
10. Spacer
11. Spindle
12. Oil seal
13. Bearing washers
14. Needle thrust bearing
15. Bushings
16. Axle
17. "O" ring
18. Washer
19. Steering arm
20. Cotter pin
21. Castelated nut
22. Washer
23. Washer
24. Shim
25. Shim
26. Bushing
27. Pivot shaft

Axle end play should not exceed 0.20 mm (0.008 inch). If end play is excessive, renew thrust washers (23) and/or add shims (24) as required.

When renewing spindle bushings (15), the top bushing should be pressed into bore until bushing is 4.7 mm (3/16 inch) below top surface of axle on models equipped with an "O" ring (17 – Figs. 1 and 2) at top of spindle (11). On models equipped with a lip type seal (12 – Figs. 3, 4, 5 and 6) at top of spindle, top bushing should be pressed into bore until top of bushing is 7 mm (9/32 inch) below top surface of axle. Install seal with lip facing upward.

Front wheel bearings should be removed, cleaned, inspected, renewed if damaged and packed with a good quality No. 2 EP lithium base grease after each 600 hours of operation. Tighten wheel bearing retaining nut (2) until slight drag is noticed while rotating wheel hub, then loosen nut to first castellation and install cotter pin.

FRONT AXLE (FOUR WHEEL DRIVE)

2. The front axle of four wheel drive models includes the differential assembly, axle housings, drive shafts, universal joints and final drives. Refer to appropriate paragraphs 3 through 12 for service to components.

Tie rod length should be adjusted to provide front wheel toe-in of 0-5 mm (0-3/16 inch) on all models.

Models 1100-1200-1300-1500-1700-1900 So Equipped

3. **REMOVE AND REINSTALL.** To remove the complete front drive axle assembly, first raise front of tractor and

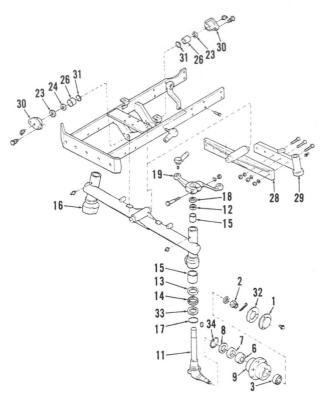

Fig. 3 — Exploded view of fixed tread front axle (16) used on 1300 and 1500 models. Axle center member (28) and extensions (29) are used on 1300, 1310, 1500 and 1510 adjustable axle models.

1. Cover
2. Castelated nut
3. Outer bearing
6. Inner bearing
7. Seal
8. Spacer
9. Hub
11. Spindle
12. Seal
13. Bearing washer
14. Needle thrust bearing
15. Bushings
16. Fixed tread axle
17. "O" ring
18. Shims
19. Steering arm
23. Washers
24. Shim
26. Bushing
28. Axle center member
29. Axle extension
30. Pivot casting
31. "O" ring
32. Gasket
33. Washer
34. Snap ring

Fig. 4 — Exploded view of adjustable front axle used on 1700, 1710, 1900 and 1910 models with two wheel drive. Refer to Fig. 3 for legend except for the following:

35. Spacer
36. Spacer
37. Retainer

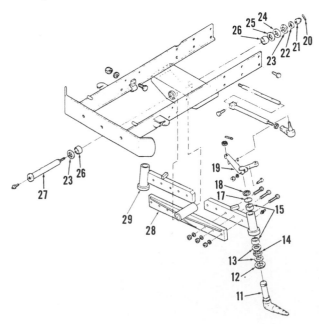

Fig. 2 — Exploded view of adjustable tread width front axle available on 1100, 1110, 1200 and 1210 models. Refer to Fig. 1 for legend except for the following:

28. Axle center member
29. Axle extensions

place a support behind the axle. Detach drag link from steering arm. Place front wheel drive control lever in "disengaged" position. Loosen clamps that attach drive shaft cover at the rear. Support axle to prevent tipping, then remove cap screws attaching axle pivot brackets to front support. Carefully lower axle until it can be moved forward out of drive shaft splines.

Inspect axle pivot bushings (13 and 36 – Figs. 7, 8 and 9) for wear or damage. Renew bushings if clearance

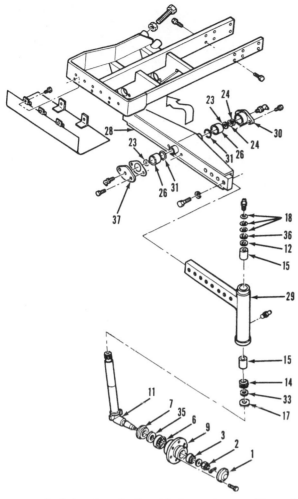

Fig. 5 — Exploded view of adjustable front axle assembly used on 1710 Offset tractors. Refer to Fig. 3 for legend except for the following:

35. Spacer
36. Spacer
37. Retainer

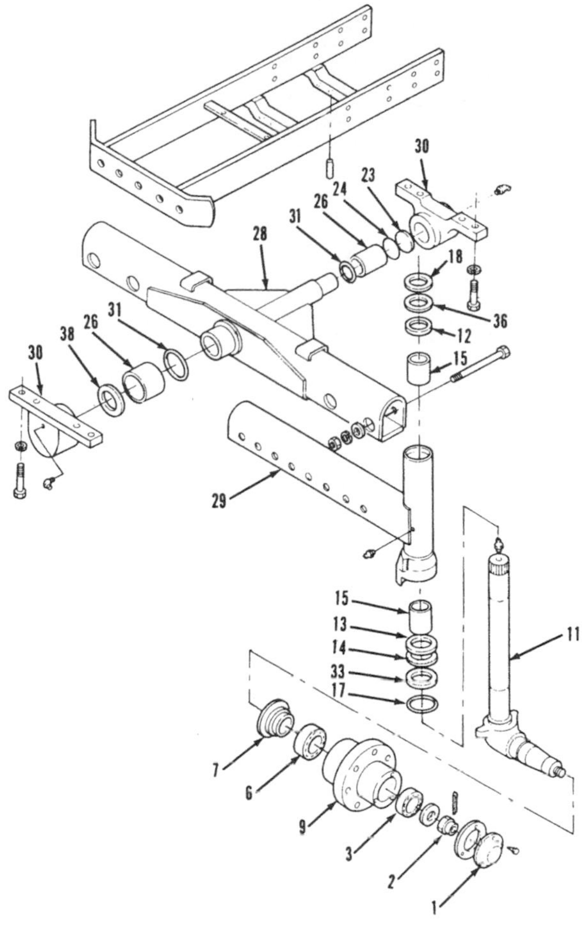

Fig. 6 — Exploded view of adjustable front axle used on Model 2110.

1. Cover	14. Needle thrust bearing	28. Axle center member
2. Nut	15. Bushings	29. Axle extension
3. Outer bearing	17. "O" ring	30. Pivot casting
6. Inner bearing	18. Shim	31. "O" ring
7. Seal	23. Washers	33. Washer
11. Spindle	24. Shim	36. Spacer
12. Seal	26. Bushings	38. Thrust washer
13. Bearing washer		

between differential case trunnions and bushings exceeds 0.35 mm (0.014 inch). Use a suitable driver to install bushings and make certain bushings are recessed 4 mm (5/32 inch) in pivot carriers to allow for installation of "O" rings (14 and 35).

When reinstalling axle, observe the following: Move axle assembly carefully into position while sliding drive shaft splines and pinion shaft splines into coupling. Tighten screws attaching pivot brackets to front support, then check axle housing fore and aft end play. Desired end play is 0.30 mm (0.012 inch) or less. If end play exceeds 0.50 mm (0.020 inch), shims should be installed in front pivot bracket as required to obtain desired end play.

Stop bolts (49 – Figs. 7, 8 and 9) should be adjusted to provide correct turning radius and to prevent drag link interference. Stop bolt setting is measured from head of bolt to surface of mounting pad. Correct length is 24 mm (15/16

inch) for 1100 and 1200 models; 32 mm (1-¼ inches) for 1300 and 1500 models; 40 mm (1-9/16 inches) for 1700 and 1900 models.

4. OUTER DRIVE ASSEMBLY. To remove the outer drive, first remove wheel and tire. Detach drag link and tie rod from axle steering arm. Remove plates and seal parts (65 through 68 – Figs. 7, 8 and 9). Support the outer drive unit, then unbolt and remove king pins (47 and 50). Withdraw assembly from axle housing (43).

On early 1100 and 1200 models, universal joint (48 – Fig. 7) is integral with shaft for outer pinion gear (53). To remove universal joint first separate outer cover (63) from housing (57) and remove snap ring (55) from end of shaft. Tap universal joint and shaft out of bearings (52 and 54) and pinion gear. Remove wheel axle (64), gear (59), bearings and seals from outer cover and housing.

On late 1100 and 1200 models and all 1300, 1500, 1700 and 1900 models, pin-

ion shaft is integral with the pinion gear (53 – Figs. 8 and 9) and universal joint (48) can be removed without disassembling outer drive unit. To disassemble outer drive, remove snap ring from inner end of pinion shaft (53). Unbolt and separate outer cover (63) from housing (57). Remove nut and washer (71), then tap wheel axle (64) out of bearings and gear. Remove pinion gear, bearings and seals from housing and cover.

On all models, backlash between pinion gear (53) and final drive gear (59) should be 0.20-0.40 mm (0.008-0.016 inch). If backlash exceeds 0.70 mm (0.028 inch), renew bearings or gears as required. Clearance between king pins (47 and 50) and bushings (45) should be 0.02-0.12 mm (0.001-0.005 inch). Maximum allowable clearance is 0.30 mm (0.012 inch).

To reassemble, reverse the disassembly procedure. Note that shims (70 – Fig. 8 and 9) are used on all except 1100 and 1200 models to adjust bearings (58 and 61) to zero end play.

1. "O" ring
2. Holder
3. Boot
4. Drive shaft housing
5. Snap rings
6. Universal joint
7. Drive shaft
8. "O" ring
9. Coupling
10. Pin (6 x 32 mm)
11. Seal
12. Rear carrier bracket
13. Bushing
14. "O" ring
15. Nut
16. Washer
17. Pinion bearing
18. Snap rings
19. Shims
20. Snap ring
21. Pinion gear
22. Ring gear
23. Differential carrier
24. Thrust washer
25. Spider gear
26. Spider shaft
27. Pin (5 x 40 mm)
28. Side gear
29. Thrust washer
30. Carrier cover
31. Carrier bearings
32. Shim
33. Shim
34. Center housing
35. "O" ring
36. Bushing
37. Carrier bracket
38. Plug
39. Gasket
40. Seal
41. Shaft
42. Pin
43. Housing
44. Thrust bearing assy.
45. King pin bearing
46. Bushing
47. Pin
48. Universal joint
49. Stop bolt
50. Pin & steering arm
51. Seal
52. Bearing
53. Outer pinion
54. Bearing
55. Snap ring
56. Fill plug
57. Housing
58. Bearing
59. Gear
60. Snap ring
61. Bearing
62. Seal
63. Outer cover
64. Wheel axle
65. Plate
66. Seal
67. Felt
68. Plate
69. Support assy.

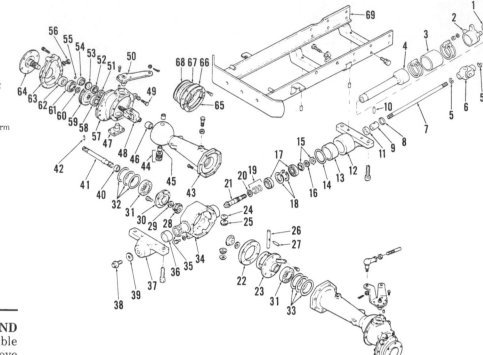

Fig. 7—Exploded view of front wheel drive axle used on 1100 and 1200 models.

5. BEVEL DRIVE GEARS AND DIFFERENTIAL.

To disassemble front axle center section, first remove both outer drive assemblies as outlined in paragraph 4. Remove front axle assembly as outlined in paragraph 3. Drain oil from axle center housing.

The differential and ring gear (22 through 30—Figs. 7, 8 and 9) can be removed after unbolting and separating axle shaft housings from center housing (34). Bevel pinion (21) can be removed after removing nuts (15) from shaft.

Retain all shims for use in reassembly. Shims (19) are used to adjust mesh of bevel pinion and ring gear. Shims (32 and 33) are used to adjust differential carrier bearing preload and bevel gear backlash.

To disassemble differential unit, unbolt and remove cover (30) from carrier (23). Remove retaining pin (27), then slide pinion shaft (26) out of carrier. Remove pinion gears (25), side gears (28) and thrust washers (24 and 29).

Backlash between differential pinion gears (25) and side gears (28) should be 0.10-0.15 mm (0.004-0.006 inch) with a wear limit of 0.50 mm (0.020 inch) for 1100, 1200, 1300 and 1500 models. Backlash between pinion gears and side

1. "O" rings
2. Holder
4. Drive shaft housing
5. Snap ring
6. Coupling
7. Drive shaft
8. "O" ring
9. Coupling
11. Seal
12. Rear carrier bracket
13. Bushing
14. "O" ring
15. Nuts
16. Washer
17. Pinion bearings
18. Snap rings
19. Shims
20. Washer
21. Pinion gear
22. Ring gear
23. Differential carrier
24. Thrust washers
25. Spider gears
26. Spider shaft
27. Pin
28. Side gear
29. Thrust washer
30. Carrier cover
31. Carrier bearings
32. Shim
33. Shim
34. Center housing
35. "O" ring
36. Bushing
37. Carrier bracket
38. Plug
39. Gasket
40. Seal
41. Shaft
43. Housing
44. Thrust bearing
45. Bushings
46. Bearing (25 x 25 mm)
47. Pin
48. Universal joint
50. Pin & steering arm
51. Seal
52. Bearing
53. Outer pinion & shaft
54. Bearing
55. Snap rings
56. Fill plug
57. Housing
58. Bearing
59. Gear
61. Bearing
62. Seal
63. Outer cover
64. Wheel axle
65. Plate
66. Seal
67. Felt
68. Plate
69. Support assy.
70. Shims
71. Nut & washer
72. Adapter plates
73. Oil seal
74. Front wheel drive housing

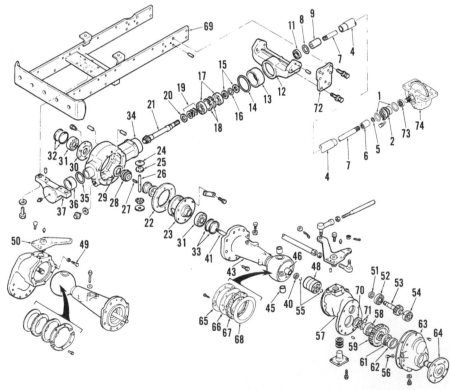

Fig. 8—Exploded view of front wheel drive axle assembly used on 1300 models.

gears for 1700 and 1900 models should be 0.05-0.10 mm (0.002-0.004 inch) with a wear limit of 0.45 mm (0.018 inch). Renew thrust washers (24 and 29) and/or gears if backlash is excessive. Diametral clearance between pinion gears (25) and shaft (26) should be 0.10-0.30 mm (0.004-0.012 inch).

Ring gear (22) and pinion (21) must be renewed as a matched set. Cap screws attaching ring gear to differential carrier (23) should be tightened to the following torque: 30-34 N·m (22-25 ft.-lbs.) on 1100 and 1200 models; 30-40 N·m (22-30 ft.-lbs.) on 1300 models; 60-70 N·m (44-51 ft.-lbs.) on 1500 models; 47-55 N·m (35-40 ft.-lbs.) on 1700 and 1900 models.

Install drive pinion (21) in center housing using shims (19) that were originally installed for initial assembly. To adjust pinion bearing preload, wrap a cord around pinion shaft as shown in Fig. 10. Use a spring scale to measure pull required to rotate the shaft. Tighten inner nut (15–Fig. 7, 8 or 9) until spring scale reading is 5-6 kg (11-13 pounds) for 1100 and 1200 models; 5½-7 kg (12-13 pounds) for 1300 and 1500 models; 11-15 kg (24¼-33 pounds) for 1700 and 1900 models. Install washer (16) and tighten outer nut (15), then recheck rolling torque.

If differential carrier (23), cover (30), carrier bearings (31), ring gear and drive pinion, center housing (34) or axle shaft housings (43) were renewed, differential carrier bearing preload, ring gear to pinion backlash and gear mesh must be checked and adjusted as outlined in paragraphs 6 and 7. If none of these components are being renewed, reassemble differential and front axle installing original shims in their original locations.

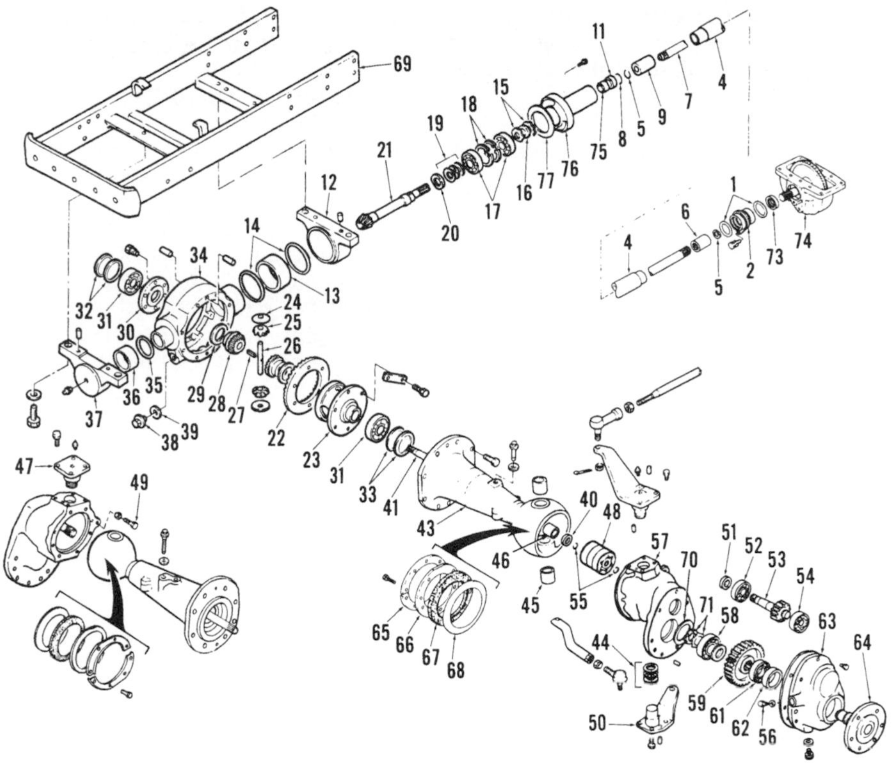

6. DIFFERENTIAL CARRIER BEARING PRELOAD. To adjust carrier bearings, first attach right axle housing to center housing (34–Fig. 7, 8 or 9). Place housing in vertical position with center housing up. Assemble sufficient thickness of shims (32) in housing bore to make sure that ring gear will not contact drive pinion, then install differential assembly in center housing. Be sure that carrier bearing is properly seated in axle housing bore.

Position left axle housing over differential assembly using more shims (33) than will be required to ensure that there is clearance between axle housing and center housing. Install four equally spaced bolts around axle housing and tighten finger tight. Use a feeler gage to measure gap between the two housings, then remove left axle housing and subtract shims from shim pack (33) equal to the measured gap.

This will provide correct preload for differential carrier bearings. Adjust ring gear to pinion backlash as outlined in paragraph 7.

Fig. 9 – Exploded view of front wheel drive axle assembly typical of type used on 1500 and 1700 models. The front wheel drive axle used on 1900 models is similar.

1. "O" ring	22. Ring gear	40. Seal	59. Gear
2. Holder	23. Differential carrier	41. Shaft	61. Bearing
4. Drive shaft housing	24. Thrust washer	43. Housing	62. Seal
5. Snap ring	25. Spider gear	44. Thrust bearing	63. Outer cover
6. Coupling	26. Spider shaft	45. Bushings	64. Wheel axle
7. Drive shaft	27. Pin	46. Bushings	65. Plate
8. "O" ring	28. Side gear	47. Pin	66. Seal
9. Coupling	29. Thrust washer	48. Universal joint	67. Felt
11. Seal	30. Carrier cover	50. Pin & steering arm	68. Plate
12. Rear carrier bracket	31. Carrier bearings	51. Seal	69. Support assy.
13. Bushing	32. Shim	52. Bearing	70. Shims
14. "O" rings	33. Shim	53. Outer pinion &	71. Nut & washer
15. Nuts	34. Center housing	shaft	73. Oil seal
16. Washer	35. "O" ring	54. Bearing	74. Front wheel drive
17. Pinion bearings	36. Bushing	55. Snap rings	housing
18. Snap rings	37. Carrier bracket	56. Fill plug	75. Bearing
19. Shims	38. Plug	57. Housing	76. Casting
20. Washer	39. Gasket	58. Bearing	77. Gasket
21. Pinion gear			

7. RING GEAR TO PINION BACKLASH. The backlash between ring gear and pinion should be 0.10-0.15 mm (0.004-0.006 inch). With left axle housing removed, backlash can be checked using a dial indicator as shown in Fig. 11. To adjust backlash, move shims (32) from right axle housing to left axle housing.

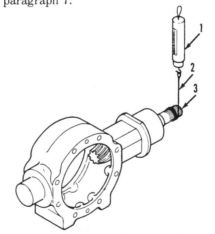

Fig. 10 – Wrap a cord (2) around pinion shaft (3) and use a spring scale (1) to check pinion rolling torque. Refer to text for adjustment.

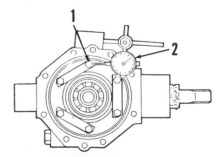

Fig. 11 – Use a dial indicator (2) to measure ring gear (1) backlash. Refer to text for adjustment.

NOTE: Do not increase or decrease overall shim pack thickness as carrier bearing preload adjustment would be affected.

Apply Prussian Blue to pinion gear teeth. Rotate pinion and note gear teeth markings made on ring gear as shown in Fig. 12. If necessary, adjust pinion gear position by changing thickness of shims (19 – Fig. 7, 8 or 9) to obtain desired tooth contact. Note that if pinion gear is moved, ring gear must also be moved to maintain recommended backlash.

Models 1110-1210-1310-1510-1710-1910-2110 So Equipped

8. **REMOVE AND REINSTALL.** To remove the complete front wheel drive axle assembly, first raise front of tractor and place a suitable support behind the axle. Detach drag link from steering arm. Place front wheel drive control lever in "disengaged" position. Loosen clamps that attach drive shaft cover (if used) at the rear. Support axle to prevent tipping, then remove cap screws attaching axle pivot brackets to front support. Carefully lower axle until it can be moved forward out of drive shaft splines.

Inspect axle pivot bushings for wear or damage. Renew bushings if clearance between differential case trunnions and bushings exceeds 0.35 mm (0.014 inch). Use a suitable bushing driver to install bushings. Be sure that bushings are recessed into axle pivot brackets far enough to allow for installation of "O" rings or seals in the pivot brackets.

To reinstall axle assembly, reverse the removal procedure. Check axle housing fore and aft end play. Desired end play is 0.30 mm (0.012 inch) or less. If end play exceeds 0.50 mm (0.020 inch), shims (4 – Fig. 17, 18, 19 or 20) should be installed in front pivot bracket (1) as required to obtain desired end play.

9. **OUTER DRIVE ASSEMBLY.** Refer to appropriate Fig. 13, 14 or 15 for an exploded view of outer drive assembly.

To remove outer drive as a unit, first support front of tractor and remove front wheel and tire. Drain oil from outer drive housing and axle center housing. Disconnect tie rod and drag link from steering arms. Detach steering dampener cylinder (if so equipped). Support the outer drive unit, then remove cap screws attaching pinion gear case (14) to axle housing and withdraw outer drive assembly.

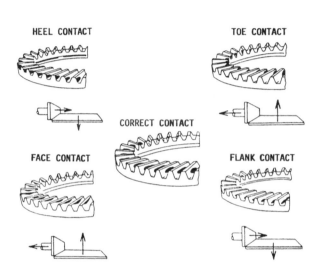

Fig. 12 – Drawings showing examples of drive pinion to ring gear tooth contact. Desired tooth contact pattern is shown in center drawing.

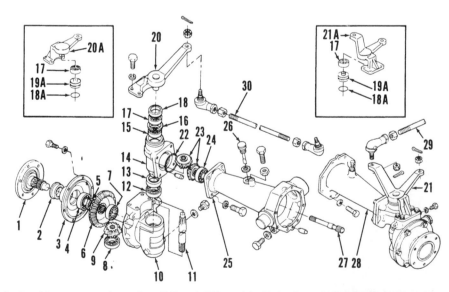

Fig. 13—Exploded view of outer drive components used on 1110 and 1210 models. Parts shown in inserts are used on late models effective with the following serial numbers: SN UB00606 for 1110 and SN UC0105 for 1210.

1. Wheel axle	9. Pinion gear	15. Bearing
2. Oil seal	10. Reduction gear housing	16. Snap ring
3. Outer cover	11. Drive shaft	17. Bearing
4. Bearing	12. Bearing	18. Oil seal
5. Snap ring	13. Oil seal	18A. "O" ring
6. Bevel gear	14. Pinion gear case	19A. Kingpin
7. Bearing		
8. Bearing		

20. Steering arm & kingpin	24. Bearing
20A. Steering arm	25. Axle housing (R.H.) & differential housing assy.
21. Steering arm & kingpin	26. Dipstick
21A. Steering arm	27. Axle shaft
22. "O" ring	28. Axle housing (L.H.)
23. Pinion gears	29. Drag link
	30. Tie rod

It is suggested that backlash between bevel gear (6) and pinion gear (9) be checked prior to disassembly to determine if gears or bearings are excessively worn. Thread a long bolt (1 – Fig. 16) into the oil drain plug hole until it contacts pinion gear to prevent the gear from moving. Install a bolt (3) in wheel hub flange (2) and position a dial indicator (4) against head of bolt shown. Rotate hub back and forth and note dial indicator reading. Normal backlash is 0.20-0.40 mm (0.008-0.016 inch). If backlash exceeds 0.50 mm (0.020 inch), renew bearings or gears as necessary.

To disassemble outer drive, remove outer cover retaining cap screws and remove cover (3 – Fig. 13, 14 or 15) with wheel axle (1) and bevel gear (6) as as assembly. Remove inner bearing (7), bevel gear (6) and snap ring (5) from wheel axle shaft, then drive the axle shaft out of the outer cover. Remove cap screws attaching steering arm (20 or 21) to reduction gear housing (10) and separate housing from pinion gear case (14).

On early 1310, 1510 and 1710 models, remove cap screw attaching steering arm to king pin (19 – Fig. 14). On all models, remove bearings, pinion gears (9 and 23) and drive shaft (11) from the housings.

Inspect all parts for wear or damage and renew as necessary. Renew all "O" rings and oil seals when reassembling.

10. BEVEL DRIVE GEARS AND DIFFERENTIAL. To remove differential and bevel gears, first remove both outer drive assemblies as outlined in paragraph 9. Remove front axle assembly as outlined in paragraph 8. The differential and ring gear can be removed after unbolting and separating left-hand axle housing (1110 and 1210 models) or right-hand axle housing (all other models) from differential center housing (19 – Fig. 17, 18, 19 or 20). Bevel pinion (20) can be removed after removing nuts (25) from the shaft. Drive the shaft inward until free of rear bearing (23). Be sure to note the location and

thickness of all shims and retain for use in reassembly.

To disassemble differential unit on 1110 and 1210 models, remove carrier bearings (9 – Fig. 17) and side gears (14). Drive out the retaining pin (11) and remove pinion gear shaft (10), pinion gears (15) and thrust washers (16).

To disassemble differential unit on 1310 and 1510 models, remove differential case cover (7 – Fig. 19), side gear (14) and thrust washer (8). Remove pinion gear shaft retaining pin, then withdraw shaft (10), pinion gears (15), thrust washers (16) and the other side gear (14) from differential case (12).

To disassemble differential unit on 1710 and 1910 models, remove left side gear (14 – Fig. 18) and bearing (9). Remove snap ring (11), then slide pinion gear shaft (10) out of differential case. Remove pinion gears (15), thrust washers (16) and right side gear (14).

To disassemble differential unit on 2110 models, remove cover (7 – Fig. 20), side gear (14) and thrust washer (8) from

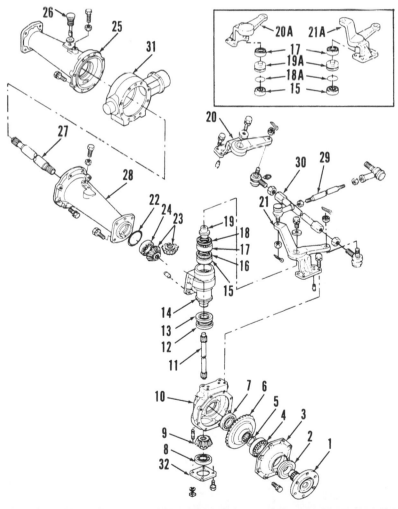

Fig. 14—Exploded view of outer drive assembly used on 1310 and 1510 models. Drive assembly used on 1710 and 1910 models is similar. Parts shown in inset are used on late production tractors. Refer to Fig. 13 for legend except for differential housing (31) and cover plate (32).

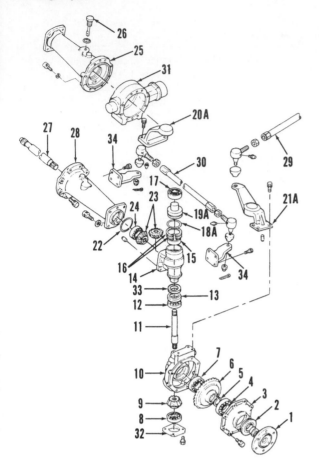

Fig. 15 — Exploded view of outer drive assembly used on 2110 models.

1. Wheel axle
2. Oil seal
3. Outer cover
4. Bearing
5. Snap ring
6. Bevel gear
7. Bearing
8. Bearing
9. Pinion gear
10. Reduction gear housing
11. Drive shaft
12. Bearing
13. Seal
14. Pinion gear case
15. Bearing
16. Snap rings
17. Bearing
18A. "O" ring
19A. Kingpin
20A. Arm
21A. Steering arm
22. "O" ring
23. Pinion gears
24. Bearing
25. Axle housing (R.H.)
26. Dipstick
27. Axle shaft
28. Axle housing (L.H.)
29. Drag link
30. Tie rod
31. Differential housing
32. Cover plate
33. Oil seal
34. Tie rod arms

Models 1710-1910
New Bearings . . . . .11-15 kg
(24.5-33 lbs.)
Used Bearings . . . . .5.5-7.5 kg
(12-16.5 lbs.)
Model 2110
New Bearings . . . . .15-20 kg
(33-44 lbs.)
Used Bearings . . . . .7.5-10 kg
(16.5-22 lbs.)

Install lockwasher (26) and tighten outer nut securely, then recheck rolling torque. Bend tabs of washer into slots of nuts when preload adjustment is correct.

If differential case (12), carrier bearings (9), axle bearings, axle housings or differential center housing (19) were renewed, differential carrier bearing preload must be adjusted as outlined in paragraph 11. Bevel ring gear to pinion backlash and gear mesh should be checked and adjusted as outlined in paragraph 12. If none of these components are being renewed, reassemble differential and front axle installing original shims in their original locations.

11. DIFFERENTIAL CARRIER BEARING PRELOAD. To adjust carrier bearing preload, proceed as follows:
On 1110 and 1210 models, assemble axle shaft (18 – Fig. 17), shims (17), bearing (7), spacer (8) and differential assembly into right axle housing (19). Support the axle housing in the upright position with differential up. Make certain there is clearance between ring gear and bevel pinion. Add shims (17) if necessary. Assemble left axle shaft (5), inner bearing and spacer with more shims (6) than necessary into left axle housing. Position the axle housing assembly onto differential housing making sure there is clearance between the two housings. Use a feeler gage to measure the gap between the housings, then remove left axle housing and subtract shims from shim pack (6) equal to the measured gap.

differential case. Remove pinion gear shafts (10 and 10A), pinion gears (15), thrust washers (16) and pinion shaft support (11). Remove the other side gear (14) and thrust washer (8).

Backlash between differential pinion gears and side gears should be 0.10-0.15 mm (0.004-0.006 inch) on 1110, 1210, 1310 and 1510 models and 0.05-0.10 mm (0.002-0.004 inch on 1710, 1910 and 2110 models. Renew thrust washers and/or gears if backlash is excessive. Clearance between pinion gear shaft and pinion gears should be 0.10-0.30 mm (0.004-0.012 inch). Renew shaft and/or gears if clearance exceeds 0.50 mm (0.020 inch).

Bevel ring gear (13) and pinion (20) must be renewed as a matched set on all models. Cap screws attaching ring gear to differential case should be tightened to 30-34 N·m (22-25 ft.-lbs.) torque on 1110 and 1210 models and to 47-55 N·m (35-40 ft.-lbs.) torque on all other models.

Install bevel pinion (20 – Fig. 17, 18, 19 or 20) in differential housing (19) using shims (22) that were originally installed. To adjust pinion bearing preload, wrap a cord around pinion shaft as shown in Fig. 21. Use a spring scale to measure pull required to rotate the

shaft. Tighten inner nut (25 – Fig. 17, 18, 19 or 20) until the specified spring scale reading is obtained as listed below.

Models 1110-1210
New Bearings . . . . .5-6 kg
(11-13 lbs.)
Used Bearings . . . . .2.5-3 kg
(5.5-6.5 lbs.)
Models 1310-1510
New Bearings . . . . .7-9 kg
(15.5-20 lbs.)
Used Bearings . . . . .3.5-4.5 kg
(8-10 lbs.)

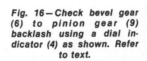

Fig. 16 — Check bevel gear (6) to pinion gear (9) backlash using a dial indicator (4) as shown. Refer to text.

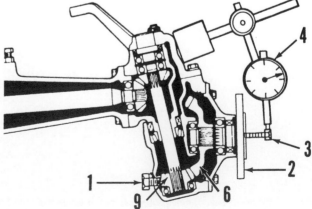

1. Pivot bracket
2. Bushing
3. "O" ring
4. Shim
5. Axle shaft, L.H.
6. Shims
7. Bearing
8. Spacer
9. Bearing
10. Pinion gear shaft
11. Pin
12. Differential case
13. Ring gear
14. Side gears
15. Pinion gears
16. Thrust washers
17. Shims
18. Axle shaft, R.H.
19. Differential housing
20. Bevel pinion
21. Thrust washer
22. Shims
23. Bearings
24. Snap rings
25. Nuts
26. Lockwasher
27. Oil seal
28. "O" ring
30. Bushing
31. Pivot bracket
32. Dowel
33. "O" ring
34. Coupling
35. Pin
36. Drive shaft
37. Tube
38. Clamps
39. Rubber cover
40. Snap rings
41. Universal joint
42. Cover
50. Support assy.

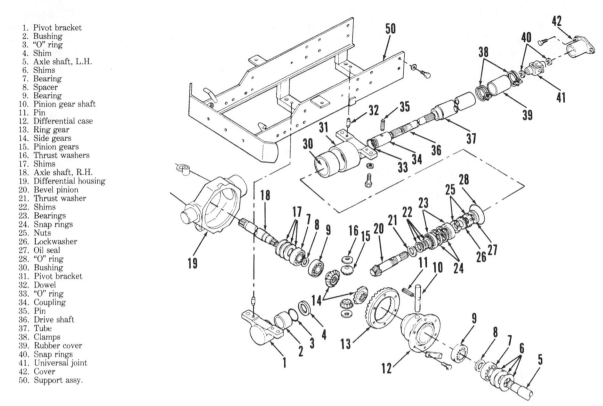

Fig. 17 — Exploded view of front wheel drive differential assembly and bevel drive gears used on 1110 and 1210 models.

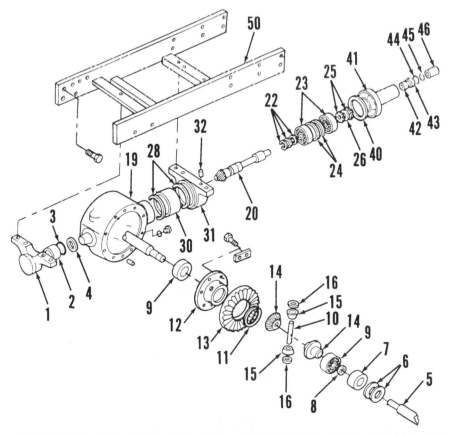

11. Snap ring
40. Gasket
41. Housing
42. Bearing
43. Oil seal
44. "O" ring
45. Snap ring
46. Coupling

Fig. 18—Exploded view of front wheel drive differential assembly and bevel drive gears used on 1710 and 1910 models. Refer to Fig. 17 for legend except for those listed.

15

This will provide recommended bearing preload adjustment.

On 1310, 1510, 1710, 1910 and 2110 models, attach left axle housing to differential center housing. Note that axle shaft (5 – Fig. 18), bearing (7) and spacer (8) must first be installed in axle housing on 1710 and 1910 models. Position axle housing vertically with differential housing up, making sure that the assembly is supported by the axle housing and not the axle shaft. Install differential assembly into differential housing. Be sure that carrier bearing is correctly located in axle housing and that there is clearance between ring gear and pinion. Add shims (6—Fig. 18, 19 or 20) if necessary. Position right axle housing (with axle shaft, inner bearing and spacer on 1710 and 1910 models) over differential assembly using more shims (17) than will

be required to insure that there is clearance between axle housing and differential housing. Use a feeler gage to measure the gap between the two housings, then remove right axle housing and subtract shims from shim pack (17) equal to the measured gap. This will provide recommended bearing preload adjustment. Adjust ring gear to pinion backlash as outlined in paragraph 12.

12. RING GEAR TO PINION BACKLASH. Carrier bearing preload should be correctly adjusted as outlined in paragraph 11 before proceeding with backlash adjustment. To check and adjust backlash, proceed as follows:

On 1110 and 1210 models, backlash is checked with left axle housing removed and right axle housing and differential completely assembled. On 1310, 1510,

1710, 1910 and 2110 models, backlash is checked with right axle housing removed and left axle housing attached to differential housing. Axle shaft (5—Fig. 18), shims (16), inner bearing (7) and spacer (8) must be assembled in axle housing on 1710 and 1910 models.

On all models, support axle housing in a vertical position with differential up. Use a dial indicator to measure backlash as shown in Fig. 22. Recommended backlash is 0.10-0.15 mm (0.004-0.006 inch). To adjust backlash, move shims between the axle housings until correct backlash is obtained.

NOTE: Do not increase or decrease overall thickness of shim pack as carrier bearing preload adjustment would be affected.

1. Pivot bracket
2. Bushing
3. Seal
4. Shim
6. Shims
7. Cover
8. Thrust washers
9. Bearing
10. Pinion gear shaft
11. Snap ring
12. Differential case
13. Ring gear
14. Side gears
15. Pinion gears
16. Thrust washers
17. Shims
19. Differential housing
20. Bevel pinion
21. Thrust washer
22. Shims
23. Bearing
24. Snap rings
25. Nuts
26. Lockwasher
28. Seal
30. Bushing
31. Pivot bracket
32. Dowel
35. Bearing
36. Oil seal
37. "O" ring
38. Collar
39. Oil seal
40. "O" ring
41. Coupling
50. Support assy.

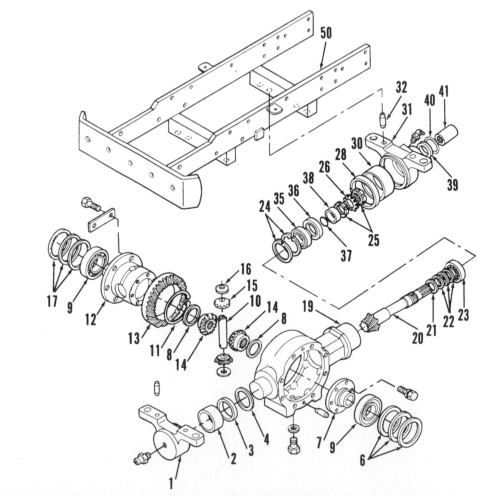

Fig. 19 — Exploded view of front wheel drive differential assembly and bevel drive gears used on 1310 and 1510 models.

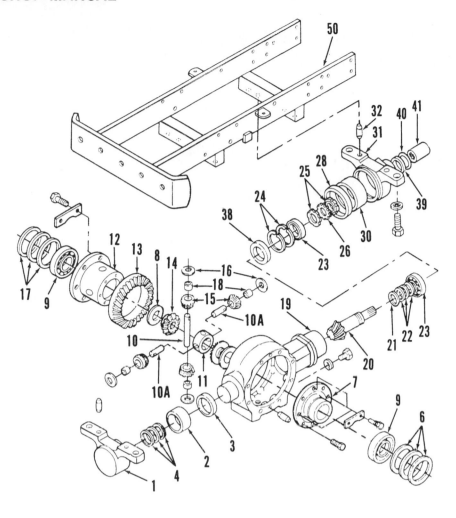

Fig. 20 — Exploded view of front wheel drive differential assembly and bevel drive gears used on 2110 models. Refer to Fig. 19 for legend except for the following:

10. Pinion gear shaft (long)
10A. Pinion gear shafts (short)
11. Pinion shaft support
18. Bushings

Fig. 22 — Use a dial indicator (2) to measure bevel ring gear (1) backlash. Refer to text for adjustment.

Apply Prussian Blue to pinion gear teeth. Rotate pinion and note gear contact markings made on ring gear. Compare markings with those shown in Fig. 23 and adjust pinion gear position, if necessary, by changing thickness of shims (22 – Fig. 17, 18, 19 or 20) to obtain desired tooth contact. Note that if pinion gear is moved, ring gear must also be moved to maintain recommended backlash.

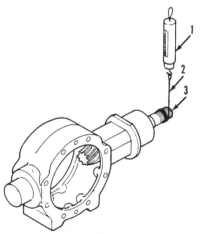

Fig. 21 — Wrap a cord (2) around bevel pinion shaft (3) and use a spring scale (1) to check pinion rolling torque. Refer to text for adjustment.

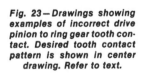

Fig. 23 — Drawings showing examples of incorrect drive pinion to ring gear tooth contact. Desired tooth contact pattern is shown in center drawing. Refer to text.

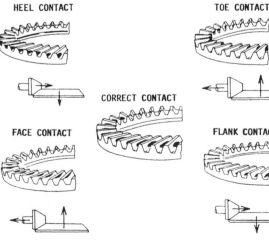

HEEL CONTACT

TOE CONTACT

CORRECT CONTACT

FACE CONTACT

FLANK CONTACT

Fig. 24—To adjust steering wheel free play (P) on models with manual steering, loosen locknut (1) and turn adjusting screw (2).

MANUAL STEERING GEAR

All Models So Equipped (Except 1710 Offset)

15. All manual steering models (except 1710 Offset) are equipped with a recirculating ball nut and sector shaft type steering gear as shown in Fig. 25, 26, 27 or 28. Steering gear is mounted on top of clutch housing and can be removed after removing steering wheel and instrument panel. On some models the fuel tank must also be removed.

Steering wheel free play (P—Fig. 24) can be externally adjusted. Free play should be 20-30 mm (¾-1⅛ inches) measured at outer rim of steering wheel. If free play is excessive, loosen locknut (1) and turn adjusting screw (2) clockwise to reduce free play. Tighten locknut to secure adjusting screw after desired setting is obtained.

16. **OVERHAUL.** To disassemble removed steering gear, proceed as follows: Scribe alignment marks on pitman arm (12—Fig. 30) and on end of sector shaft (7), then use a suitable puller to remove pitman arm. Remove retaining screws (3—Fig. 25, 26, 27 or 28), then withdraw sector shaft (7) and cover (4) as an assembly. Remove screws attaching steering column cover (21) to housing (8), then pull worm shaft, ball nut (26)

Fig. 26—Exploded view of manual steering gear used on 1300, 1310, 1510 and 1710 models. Refer to Fig. 25 for legend except for the following:

33. Steering shaft
34. Pins
35. Spring
36. "O" rings
37. Worm shaft
38. Bushing

and related parts from housing. Note that worm shaft and ball nut are available only as an assembly and should not be disassembled. Separate column cover from steering shaft.

Inspect all parts for wear or damage. Renew components as required if clearance between sector shaft (7) and bushings, located in side cover (4) and body (8), exceeds 0.15 mm (0.006 inch). Renew all seals and gaskets.

When reassembling, install ball nut, steering shaft and column cover using original shim (23) between steering column and housing. Attach a cord to steering shaft and measure pull required to rotate shaft using a spring scale as shown in Fig. 29. Add or remove shims (23—Fig. 25, 26, 27 or 28), if necessary, to obtain the following recommended preload setting.

Models	Spring Scale Reading
1100-1110-1200-1210	2.7-6 kg (6-13 lbs.)
1300-1310-1500-1510	1.7-4.5 kg (4-10 lbs.)
1700-1710	3.1-6.9 kg (7-15 lbs.)
1900-1910-2110	2.5-7.7 kg (5.5-17 lbs.)

Locate ball nut (26—Fig. 25, 26, 27 or 28) in center of travel on worm shaft. Install sector gear and shaft (7) with center tooth of sector gear engaging center teeth of ball nut. Tighten side cover (4) retaining screws.

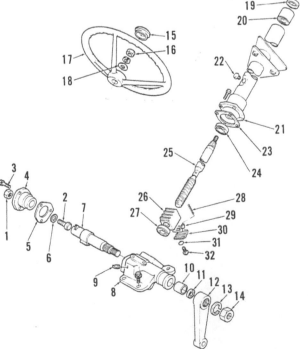

Fig. 25—Exploded view of manual steering gear used on 1100, 1110, 1200 and 1210 models.

1. Locknut
2. Adjusting screw
3. Cap screw
4. Cover
5. Gasket
6. Shim
7. Sector shaft
8. Housing
9. Plug
10. Bushing
11. Seal
12. Pitman arm
13. Washer
14. Nut
15. Cap
16. Nut
17. Steering wheel
18. Washer
19. Seal
20. Bushing
21. Cover
22. Plug
23. Shim
24. Bearing
25. Shaft
26. Ball nut
27. Bearing
28. Balls
29. Ball guides
30. Retainer
31. Washer
32. Screws

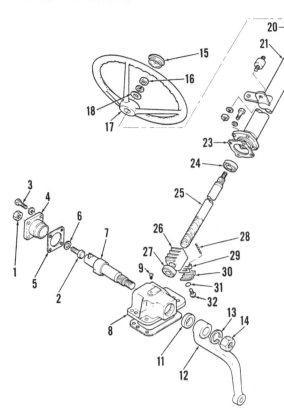

Fig. 27 — Exploded view of manual steering gear used on 1500 and 1700 models.

1. Locknut
2. Adjusting screw
3. Cap screw
4. Cover
5. Gasket
6. Shim
7. Sector shaft
8. Housing
9. Plug
11. Seal
12. Pitman arm
13. Washer
14. Nut
15. Cap
16. Nut
17. Steering wheel
18. Washer
20. Bushing
21. Cover
23. Shim
24. Bearing
25. Shaft
26. Ball nut
27. Bearing
28. Balls
29. Ball guides
30. Retainer
31. Washer
32. Screw

then pull ball nut assembly (15) from gear housing.

Inspect all parts for wear or damage. Maximum allowable clearance between sector shaft (8) and bushings (5 and 18) is 0.15 mm (0.006 inch). Renew "O" ring, gasket and seals.

When reassembling, install ball nut assembly and rear cover in gear housing. Wrap a cord around ball nut shaft and use a spring scale to measure pull required to rotate the shaft. Thread rear cover (12) into housing to preload bearings until spring scale reading is 6.7-8.7 kg (15-19 lbs.) When correct preload is obtained, install locknut (10) and tighten securely.

Assemble adjusting screw (7), shim (6) and top cover (3) to the sector shaft. Locate ball nut in center of travel on worm shaft. Install sector gear and shaft with center tooth of sector gear engaging center teeth of ball nut. Tighten side cover retaining screws. In-

Fill steering gear to level plug opening (9) with Ford 134 hydraulic fluid or equivalent. Install pitman arm aligning assembly marks (Fig. 30). Install steering wheel and adjust free play as outlined in paragraph 15.

Model 1710 Offset

17. The 1710 Offset model is equipped with manual steering as standard equipment. (Refer to paragraph 22 for power steering equipped tractors). The steering gear is mounted on the front axle support frame.

Steering wheel free play can be externally adjusted. Desired free play is 20-30 mm (³⁄₄-1¹⁄₈ inches) measured at outer rim of steering wheel. If free play is excessive, loosen adjusting screw locknut (1 – Fig. 31) and turn adjusting screw (7) clockwise to reduce free play. Tighten locknut while holding adjusting screw after desired setting is obtained.

18. **OVERHAUL.** When removing steering gear, be sure to scribe reference marks on pitman arm (20 – Fig. 31) and sector shaft (8) to ensure correct alignment for reassembly. Disconnect pitman arm and steering shaft universal joint from steering gear, remove mounting bolts and withdraw steering gear.

To disassemble, remove adjusting screw nut (1). Remove cap screws retaining top cover (3), then remove cover and sector shaft (8) as an assembly. Remove nut (10) and rear cover (12),

Fig. 28 — Exploded view of manual steering gear used on 1900 models. Steering gear used on 1910 and 2110 manual steering models is similar. Refer to Fig. 27 for legend except for the following:

10. Bushing
38. Bushing
39. Plate
40. Steering column

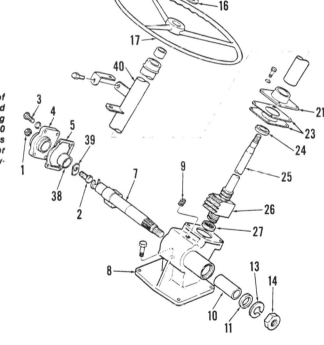

Fig. 29 — To check steering shaft bearing preload, use a cord and spring scale to measure pull required to rotate shaft. Shims located between steering column and housing are used to adjust preload. Refer to text.

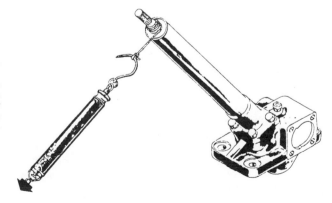

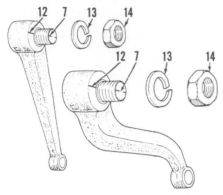

Fig. 30—Reference marks should be scribed on end of sector shaft (7) and pitman arm (12) to ensure correct alignment when reassembling.

stall pitman arm on sector shaft being sure to align reference marks made during disassembly. If no alignment marks are present, locate ball nut in center of travel on worm shaft, then install pitman so it points straight forward.

Install steering gear assembly, then turn adjusting screw (7) to adjust clearance between sector gear and ball nut so steering wheel free play is 20-30 mm (¾-1⅛ inches). Install seal washer (2) and tighten locknut (1) to 30-40 N·m (22-30 ft.-lbs.) torque.

POWER STEERING SYSTEM

All Models So Equipped

20. An integral hydraulic power assist steering gear is available on some models. On all models except 2110, a priority valve is used to divert part of the flow from the hydraulic lift system pump to operate the power steering. On 2110 models, a separate hydraulic pump supplies fluid to steering gear assembly. The pump is located on left side of engine and driven by the camshaft gear.

On all models, steering wheel free play can be adjusted externally. Desired free play is 20-30 mm (¾-1⅛ inches) measured at outer rim of steering wheel. If free play is excessive, adjust sector shaft to ball nut clearance as follows: Start engine and run at idle speed. Loosen locknut (39—Fig. 32) and turn adjusting screw (45) clockwise to reduce free play to desired setting. Tighten locknut after adjustment is completed.

All Models So Equipped (Except 1710 Offset)

21. **POWER STEERING GEAR.** The steering gear assembly can be removed after first removing the fuel tank, steering wheel and instrument panel. Scribe reference marks on pitman arm and end of sector shaft (Fig. 34), then use a suitable puller to remove pitman arm from shaft.

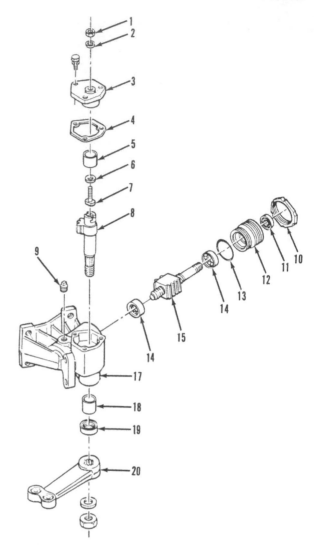

Fig. 31—Exploded view of steering gear assembly used on manual steering 1710 Offset models.

1. Nut
2. Seal washer
3. Cover
4. Gasket
5. Bushing
6. Shim
7. Adjusting screw
8. Sector shaft & gear
9. Plug
10. Nut
11. Oil seal
12. Rear cover
13. "O" ring
14. Bearings
15. Ball nut assy.
17. Housing
18. Bushing
19. Oil seal
20. Pitman arm

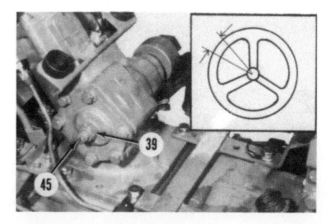

Fig. 32—To adjust steering wheel free travel, loosen locknut (39) and turn adjusting screw (45) as required.

Before disassembling steering gear, an assembly plate should be fabricated using the dimensions shown in Fig. 35. The assembly plate will be used during assembly of steering shaft and adjustment of bearing preload.

Scribe assembly reference marks on steering column (36—Fig. 33) and seal retainer housing (28). Loosen nut (35) and remove steering column. Remove rivet (32) and spring pin (33) from steer-ing shaft (34), then remove shaft, spring (31) and "O" rings (14).

Remove seal retainer housing (28) and "O" ring. Install the special assembly plate (Fig. 35) on the control valve housing to hold control valve centering pistons (23—Fig. 33) and springs in place as the retaining nut (26) is being removed. Remove assembly plate and control valve housing (19) assembly being careful not to lose centering pistons and

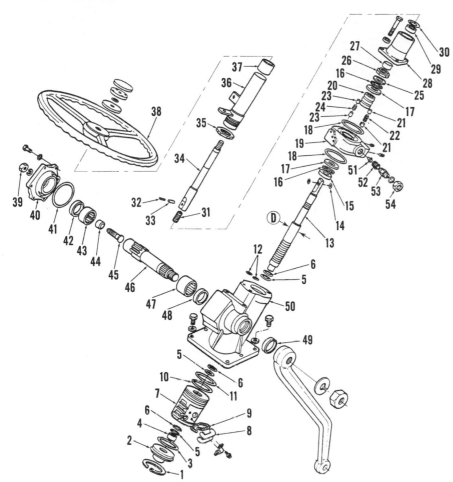

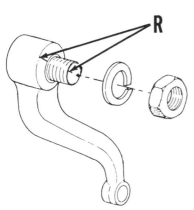

Fig. 33 — Exploded view of typical power steering gear assembly available on 1310, 1510, 1700, 1710, 1900, 1910 and 2110 models. Plungers (21) are not used on 1310, 1510, 1710, 1910 and 2110 models.

1. Snap ring
2. Cover
3. "O" ring
4. Bearing
5. "O" ring
6. Sealing rings
7. Ball nut
8. Guides
9. Balls
10. "O" ring
11. Sealing ring
12. "O" rings
13. Steering worm shaft
14. "O" rings
15. Thrust washer
16. Needle bearing
17. Thrust washer
18. "O" rings
19. Valve body
20. Valve spool
21. Active plungers (4 used)
22. Springs (2 used)
23. Inactive plungers (4 used)
24. Springs (2 used)
25. Thrust washer
26. Nut
27. Needle bearing
28. Cover
29. Oil seal
30. Snap ring
31. Spring
32. Rivet
33. Spring pin
34. Steering wheel shaft
35. Nut
36. Column
37. Bushing
38. Steering wheel
39. Locknut
40. Cover
41. "O" ring
42. Oil seal
43. Bearing
44. Adjuster nut
45. Adjuster screw
46. Sector shaft
47. Bearing
48. Oil seal
49. Seal
50. Housing
51. Poppet
52. Spring
53. Adjuster
54. Locknut

springs. Remove relief valve assembly (51 through 54) only if necessary.

Remove adjusting screw locknut (39) and the cap screws attaching cover (40) to steering gear housing (50). Turn the adjusting screw (45) clockwise to remove the cover. Turn worm shaft (13) until teeth of sector (46) are centered with opening of housing, then remove sector shaft from housing.

Remove snap ring (1) and bottom cover (2), then withdraw ball nut (7) and worm shaft (13) as an assembly through bottom opening in housing. Ball races (8), balls (9), ball nut (7) and worm shaft (13) are available only as a set and should not be disassembled.

Inspect all parts for wear or damage. Renew worm shaft and ball nut as an assembly if shaft does not rotate smoothly. Clamp ball nut in a vise and measure end play of worm shaft with a dial indicator. Renew shaft and ball nut if end play exceeds 0.10 mm (0.004 inch). Worm shaft bearing journal diameter (D) wear limit is 24.90 mm (0.980 inch).

Inspect bearing and oil seal journal surfaces of sector shaft (46) for wear and renew if necessary. Adjusting screw (45) should have 0.05-0.07 mm (0.002-0.003 inch) end play in sector shaft.

Check steering gear housing (50) cylinder bore for wear or score marks. Renew parts as necessary if clearance between ball nut and housing bore exceeds 0.15 mm (0.006 inch).

Inspect valve spool (20) and valve body (19) for wear or scratches. Edges of annular rings of spool and body should be sharp and undamaged. Clearance between spool and bore in valve body should not exceed 0.025 mm (0.001 inch).

R

Fig. 34 — Scribe reference marks (R) on pitman arm and end of sector shaft prior to removal to ensure correct reassembly.

Fig. 35—Special assembly tool should be fabricated from 10 mm (3/8 inch) thick steel or hard plastic plate. Dimension (A) is circle diameter for the three equally spaced 11 mm (7/16 inch) diameter holes which must match seal retainer housing (28—Fig. 33) mounting hole pattern.

A. 86-87 mm (3.386-3.425 in.)
B. 107 mm (4.213 in.)
C. 54 mm (2.125 in.)
D. 74.5-75 mm (2.933-2.957 in.)

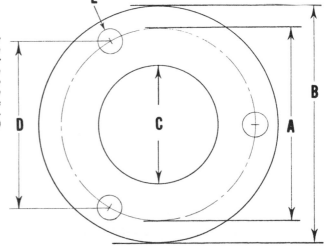

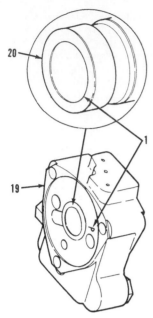

Fig. 36—Install valve spool (20) so end with shortest ID chamfer (1) is on side of valve body (19) marked "P."

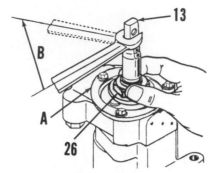

Fig. 37—Use special assembly plate (A) to hold control valve assembly in place on steering gear housing when adjusting thrust bearing preload. Refer to text for procedure.

A. Assembly plate
B. ⅛ turn (45°)
13. Worm shaft
26. Adjusting nut

When renewing worm shaft needle bearings (4 and 27), be sure numbered side of bearings face up. Install new oil seal (29) in retainer housing (28) with open side of seal facing down. When renewing sector shaft oil seals, install inner seal (48) with open side facing inward and install outer seal (49) with open side facing outward. Press against numbered side of bearing when installing sector shaft needle bearings (43 and 47). Be careful not to drive inner bearing (47) all the way in against inner seal.

Renew all "O" rings and sealing rings. Teflon sealing rings can be softened by soaking in hot water to facilitate installation. Compress Teflon ring (11) in-

to seal groove after installation until it returns to its original shape.

To reassemble, reverse the disassembly procedure while noting the following special instructions. Position ball nut in center of travel on worm shaft, then install sector shaft with center tooth of sector gear engaging center teeth of ball nut. Assemble valve spool (20–Fig. 36) in valve body (19) so short inside diameter chamfer end of spool is on side of body marked with a "P." Assemble control valve assembly and thrust bearings over worm shaft, then install special assembly plate (Fig. 35) to hold control valve components in place. Adjust steering shaft thrust bearing preload as follows: Turn steering shaft (13–Fig. 37) counterclockwise while holding adjusting nut (26) until all shaft end play is removed. Then turn shaft ⅛ turn (45°) clockwise (B) while holding nut. Bend tabs of lockwasher to prevent nut from loosening. Remove assembly plate, then install "O" ring (18–Fig. 32) and retainer housing (28).

Install steering gear assembly, then adjust sector shaft to ball nut clearance to provide recommended steering wheel free play of 20-30 mm (¾-1⅛ inches) as outlined in paragraph 20. Check power steering relief valve pressure and adjust if necessary as outlined in paragraph 23.

1. Cover plate
2. Adjusting nut
3. Washer (thin)
4. Thrust bearing
5. Washer
6. "O" ring
7. Valve spool
8. Spring (2 used)
9. Centering piston (4 used)
10. Centering spring (2 used)
11. Nut
12. Adjusting screw
13. "O" ring
14. Spring
15. Relief valve poppet
16. Control valve body
18. Washer (thick)
19. "O" ring
20. Teflon seal ring
21. "O" ring
22. Oil seal
23. Gear housing
24. Seal
25. Needle bearing
26. Sector shaft
27. Adjusting screw
28. Adjuster nut
29. Needle bearing
30. "O" ring
31. Side cover
32. Seal washer
33. Jam nut
34. Cap nut
35. Teflon ring
36. "O" ring
37. "O" ring
38. Teflon ring
39. Worm shaft
40. Ball nut
41. Balls & guide
42. Bearing
43. Teflon ring
44. "O" ring
45. "O" ring
46. End cover
47. Snap ring
48. Oil seal
49. Snap ring
50. Dust cover

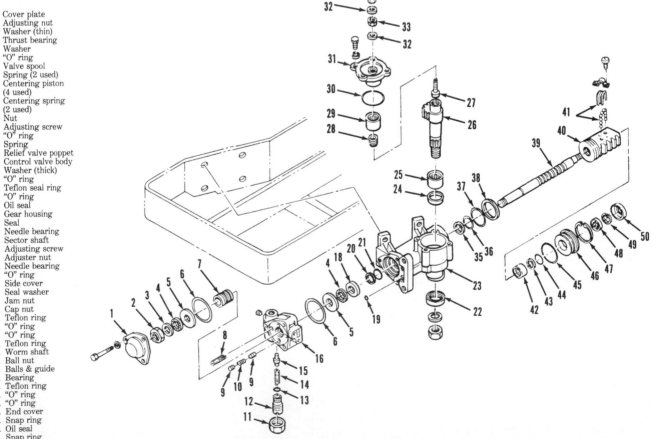

Fig. 38—Exploded view of steering gear assembly used on 1710 Offset models equipped with power steering.

Model 1710 Offset

22. POWER STEERING GEAR.
The steering gear assembly is mounted on the front axle support frame. To remove steering gear, first remove clamp bolt from steering shaft universal joint, then unbolt and remove steering wheel and shaft assembly. Scribe reference marks on pitman arm and end of sector shaft, then remove pitman arm. Disconnect hydraulic lines, remove mounting bolts and remove steering gear.

To disassemble, remove front cover (1 – Fig. 38). Remove nut (2), thrust bearing assemblies and control valve (16) assembly from worm shaft. Be careful not to lose centering pistons (9) and springs (8 and 10). Remove relief valve assembly (11 through 15) from control valve if necessary.

Remove cap nut (34) and jam nut (33) from sector shaft adjusting screw. Remove cap screws attaching side cover (31) to housing, then turn adjusting screw clockwise to force cover off the housing. Withdraw sector shaft (26) from the housing.

Remove snap ring (47), then push shaft (39), ball nut (40) and end cover (46) out of gear housing. The ball nut and shaft are available only as a set and should not be disassembled.

Inspect all parts for wear or damage. Renew ball nut assembly if worm shaft does not turn smoothly. Clamp ball nut in a vise and measure end play of shaft using a micrometer. Renew ball nut assembly if end play exceeds 0.10 mm (0.004 inch). Worm shaft bearing journal diameter wear limit is 24.90 mm (0.980 inch).

Inspect bearing and seal journal surfaces of sector shaft (26) for wear. Renew shaft if journal diameter (A – Fig. 39) is less than 31.90 mm (1.256 inches) or if diameter (B) is less than 27.90 mm (1.098 inches).

Check steering gear housing (23 – Fig. 38) cylinder bore for wear or score marks. Renew parts as necessary if clearance between ball nut and housing bore exceeds 0.15 mm (0.006 inch).

Inspect valve spool (7) and valve body (19) for wear or scratches. Renew valve assembly if clearance between spool and valve body bore exceeds 0.025 mm (0.001 inch).

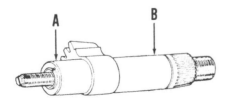

Fig. 39 – Wear limit for steering sector shaft bearing journals is 31.90 mm (1.256 inches) at (A) and 27.90 mm (1.098 inches) at (B).

To reassemble steering gear, reverse the disassembly procedure while noting the following special instructions. When renewing sector shaft seals, install inner seal (24) with open side facing inward and install outer seal (22) with open side facing outward. Install new worm shaft oil seal (48) in end cover (46) with lip facing inward. Teflon seal rings can be softened by soaking in hot water to aid installation. Be sure that end of control valve spool with the short inside diameter chamfer is on side of valve body marked with a "P" as shown in Fig. 36.

Assemble ball nut and shaft, thrust bearings and control valve assembly to gear housing. Install special assembly plate (Fig. 35) using three cap screws from cover plate to hold components in place, then adjust bearing preload as follows: While holding worm shaft, turn adjusting nut (2 – Fig. 38) until all end play is removed. Stake the nut to the shaft to prevent loosening. Remove assembly plate and install cover plate (1).

Position ball nut at center of worm shaft, then install sector shaft, centering the sector gear with the ball nut. Install side cover (31). Install pitman arm on sector shaft aligning reference marks (Fig. 34) made prior to disassembly. If

no reference marks are present, install arm so it points straight forward with steering centered between full left and full right positions.

After steering gear is reinstalled, adjust sector shaft adjusting screw (27 – Fig. 38) to provide recommended steering wheel free travel of 20-30 mm (¾-1⅛ inches). Check relief valve pressure setting as outlined in paragraph 23.

All Models So Equipped

23. RELIEF VALVE PRESSURE.
To check and adjust power steering relief valve pressure setting, install a 0-20000 kPa (0-3000 psi) pressure gage in power steering supply line using a "T" fitting. Refer to Fig. 40 or 40A. Start engine and operate steering to warm the oil. Set engine speed at 1900 rpm, turn steering wheel to full left or right position and observe pressure gage reading.

On all models except 2110, relief valve pressure should be 6900 kPa (1000 psi) on two wheel drive tractors and 10340 kPa (1500 psi) on four wheel drive tractors. On 2110 models, relief valve pressure should be 10340 kPa (1500 psi) on two wheel drive tractors and 13100 kPa (1900 psi) on four wheel drive tractors. To adjust pressure setting, loosen locknut (1 – Fig. 41 or 42) and turn ad-

Fig. 40 – Connect a pressure gage (1) into power steering supply line (2) using a tee fitting (3) to check power steering relief valve pressure setting. Priority flow control valve (4) is located on right side of hydraulic lift housing on 1700 and 1900 models and is not used on 2110 models.

Fig. 40A – A separate pump (4) is used for power steering on 2110 models. Use a tee fitting (3) to connect a gage (1) in pressure line (2) to check steering relief valve pressure setting.

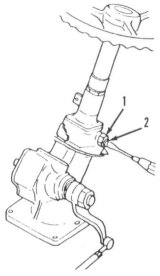

Fig. 41—Power steering relief valve adjustment point for all models except 1710 Offset.

justing screw (2) to obtain specified pressure.

All Models So Equipped

24. **PRIORITY FLOW CONTROL VALVE.** A priority flow control valve is used on all models except 2110 to ensure an adequate supply of oil is available to the power steering. The control valve divides the oil flow from hydraulic system pump to provide priority oil flow to power steering and directs remaining oil flow to tractor hydraulic system.

Fig. 44—Exploded view of priority flow control valve used on 1700 and 1900 models with power steering. Refer to Fig. 43 for legend except for shim (7).

On 1700 and 1900 models the priority valve is mounted on the right-hand side of the hydraulic hitch housing. On 1310 and 1910 models, the priority valve is attached to right-hand side frame rail. On 1510 and 1710 models, the valve is attached to left-hand side frame rail.

Refer to appropriate Fig. 43 or 44 for an exploded view of priority flow control valve. When servicing the valve, be sure orifice in spool (2) is open. Renew valve assembly if spool and body bore are excessively worn or damaged. Light scratches may be polished out with fine emery cloth.

All Models So Equipped

25. **HYDRAULIC PUMP.** Service parts are not available for repair of hydraulic pump. Renew pump assembly if it is faulty.

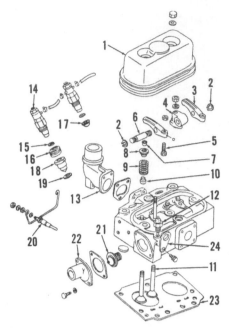

Fig. 45—Exploded view of cylinder head typical of 1100, 1110, 1200 and 1300 models.

1.	Rocker arm cover	13.	Inlet housing
2.	Snap ring	14.	Injector
3.	Rocker arm	15.	Seal
4.	Rocker arm support	16.	Screw
5.	Adjustment screw	17.	Cap
6.	Rocker arm shaft	18.	Prechamber
7.	Keepers	19.	Gasket
8.	Valve cap	20.	Glow plug
9.	Spring	21.	Thermostat
10.	Stem seal (all valves)	22.	Housing
11.	Valve	23.	Head gasket
12.	Temperature sensor	24.	Lifting eye

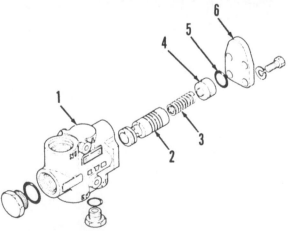

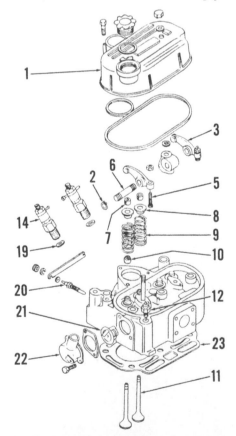

Fig. 42 — Power steering relief valve adjusting screw (2) is accessible from bottom of steering gear on 1710 Offset tractors.

Fig. 43 — Exploded view of priority flow control valve used on 1310, 1510, 1710 and 1910 models with power steering.

1. Valve body
2. Spool
3. Spring
4. Plug
5. "O" ring
6. Retainer plate

Fig. 46—Exploded view of cylinder head typical of 1500 and 1700 models. Refer to Fig. 45 for legend.

ENGINE AND COMPONENTS

R&R ENGINE ASSEMBLY

All Models

26. To remove engine assembly, first drain coolant from radiator, drain hydraulic oil from transmission housing, and drain engine oil if engine is to be disassembled. Remove front end weights if so equipped. Disconnect and remove the battery. Remove the hood assembly. Disconnect upper and lower radiator hoses and remove radiator support brace if used. Remove air cleaner hose.

Disconnect steering drag link (except 1710 Offset model) from pitman arm. On 1710 Offset, remove cap screws attaching steering column to upper support bracket and disconnect power steering (if equipped) supply and return lines. On models with hydrostatic transmission, disconnect oil lines from transmission oil cooler filter manifold.

Install wood wedges between front support frame rails and front axle to prevent tipping. Attach a hoist to engine lift brackets and place suitable support stand under clutch housing. If equipped with front wheel drive axle, place front drive control lever in disengaged position and loosen drive shaft cover clamps

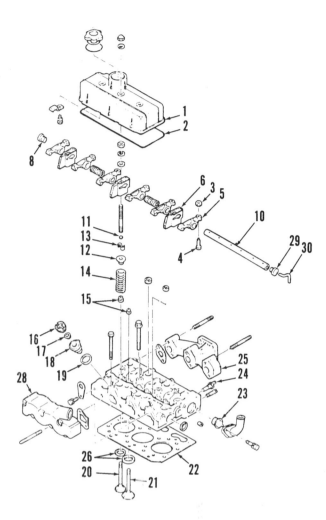

Fig. 48—Exploded view of cylinder head typical of Model 1510. Cylinder head used on Model 1710 is similar.

1. Rocker arm cover
2. Gasket
3. Locknut
4. Adjusting screw
5. Rocker arm
6. Support bracket
8. Plug
10. Rocker shaft
11. Valve stem cap
12. Retainer
13. Keeper
14. Spring
15. Valve seals
16. Retainer
17. Heat shield
18. Prechamber
19. Gasket
20. Exhaust valve
21. Intake valve
22. Head gasket
23. Thermostat
24. Temperature sender
25. Exhaust manifold
26. Valve seat inserts
28. Intake manifold
29. Plug
30. Oil tube

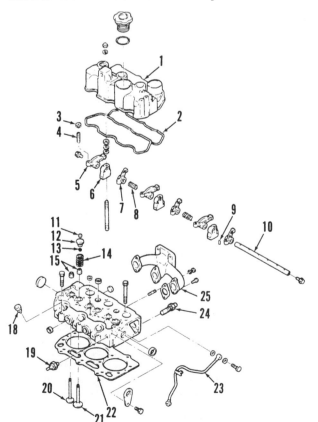

Fig. 47—Exploded view of cylinder head typical of 1210 and 1310 models.

1. Rocker arm cover
2. Gasket
3. Locknut
4. Adjusting screw
5. Rocker arm (exhaust)
6. Support bracket
7. Rocker arm (intake)
8. Spring
9. Pin
10. Rocker shaft
11. Valve stem cap
12. Retainer
13. Keeper
14. Spring
15. Valve seals
18. Prechamber
19. Oil pressure sender
20. Exhaust valve
21. Intake valve
22. Head gasket
23. Oil tube
24. Temperature sender
25. Exhaust manifold

if used. Remove cap screws attaching side frame rails to engine, then carefully roll front axle assembly away from engine.

Disconnect electrical wiring from oil pressure switch, temperature switch, glow plugs, alternator and starter motor. Remove the starter motor. Disconnect tachometer drive cable. Disconnect throttle control rod. Disconnect hydraulic system pump suction and pressure lines, power steering pressure line (if equipped) and power steering pump suction line (Model 2110). On all models, remove cap screws attaching engine to transmission and withdraw engine.

To reinstall engine, reverse the removal procedure.

CYLINDER HEAD

All Models

27. To remove cylinder head, drain coolant from radiator and cylinder block. Remove air cleaner hose, air

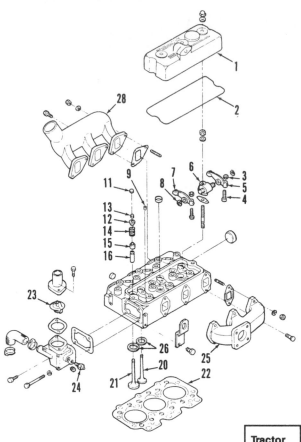

original location. Loosen cylinder head nuts or cap screws alternately a half turn at a time to prevent distortion of cylinder head, then remove head from engine.

Use a valve spring compressor and remove valve keepers, springs and valves from cylinder head. Remove glow plugs. On 1210 and 1310 models, top prechambers (18–Fig. 47) from underside of head or use a suitable puller to remove from head. On 1100, 1110, 1200, 1300, 1510 and 1710 models, unscrew retaining ring (16–Figs. 45 and 48), then tap prechamber (18) from underside of head or use a suitable puller to remove from head.

Thoroughly clean the cylinder head, then check for cracks or other damage. Use a straightedge and feeler gage to check cylinder head for flatness. Resurface or renew cylinder head if warped more than 0.12 mm (0.005 inch).

On early 1100, 1200, 1300 and 1900 models, cylinder head gasket and shims were used to adjust clearance between the top of pistons and the cylinder head. The practice of using shims was discontinued and cylinder head gaskets are

Fig. 49—Exploded view of cylinder head used on Model 1910. Model 2110 is similar except that it has a four cylinder engine. Cylinder head used on Model 1900 is similar except renewable valve guides (16) are not used. Refer to Fig. 48 for legend except for the following items.

6. Rocker shaft & support assy.
7. Rocker arm
8. Snap ring
9. Pin
16. Valve guide

cleaner (if necessary) and intake manifold. Disconnect upper radiator hose and bypass hose (if used). On 1210 and 1310 models, remove the water pump. On all other models, remove thermostat housing. On all models, remove muffler and exhaust manifold. Remove injector nozzles and immediately plug all openings in fuel system to prevent entrance of dirt. Remove the external oil line to cylinder head on 1210 and 1310 models. On all models, remove valve cover, valve rocker arm assemblies and push rods. Be sure to keep all valve components in separately marked containers so they can be reinstalled in their

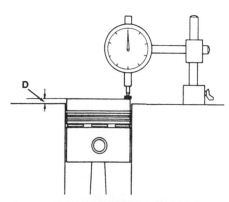

Fig. 50—Measure piston height (D) above face of cylinder block to determine proper thickness head gasket to install on 1110, 1210, 1310, 1510, 1710, 1910 and 2110 models. Refer to text and Fig. 51.

Tractor Model	Piston Height Measurement	Gasket Free Thickness	Part Number SBA-
1100	—	1.30mm (0.051 in.)	111146140
	—	1.55mm (0.061 in.)	111146621
1110	0.73-0.85mm (0.029-0.033 in.)	1.55mm (0.061 in.)	111146621
	0.85-0.99mm (0.033-0.039 in.)	1.70mm (0.067 in.)	111146622
1200, 1300	—	1.45mm (0.057 in.)	111146160
	—	1.55mm (0.061 in.)	111146631
1210	0.35-0.45mm (0.014-0.018 in.)	1.1mm (0.043 in.)	111146820
	0.45-0.55mm (0.018-0.022 in.)	1.2mm (0.047 in.)	111146830
	0.55-0.65mm (0.022-0.026 in.)	1.3mm (0.051 in.)	111146840
1310	0.45-0.55mm (0.018-0.022 in.)	1.2mm (0.047 in.)	111146880
	0.55-0.65mm (0.022-0.026 in.)	1.3mm (0.051 in.)	111146890
	0.65-0.75mm (0.026-0.030 in.)	1.4mm (0.055 in.)	111146900
1510	0.20-0.30mm (0.008-0.012 in.)	1.45mm (0.057 in.)	111146920
	0.30-0.40mm (0.012-0.016 in.)	1.55mm (0.061 in.)	111146851
	0.40-0.50mm (0.016-0.020 in.)	1.65mm (0.065 in.)	111146861
1710	0.45-0.58mm (0.018-0.023 in.)	1.45mm (0.057 in.)	111146561
	0.58-0.75mm (0.023-0.030 in.)	1.55mm (0.061 in.)	111146651
1900	—	1.3mm (0.051 in.)	111146530
	—	1.4mm (0.055 in.)	111146670
1910	Under 0.60mm (0.024 in.)	1.5mm (0.059 in.)	111146930
	Over 0.60mm (0.024 in.)	1.7mm (0.067 in.)	111146940
2110	Under 0.60mm (0.024 in.)	1.5mm (0.059 in.)	111146733
	Over 0.60mm (0.024 in.)	1.7mm (0.067 in.)	111146783

Fig. 51—Head gaskets are available in different thicknesses for some models to set piston to cylinder head clearance.

now available in two different thicknesses to set piston to cylinder head clearance. The thicker head gasket is to be used on engines that were equipped with both a gasket and shims. The thinner gasket is to be used on engines that were equipped with a gasket and no shims. For identification purposes, the last four digits of head gasket part number are stamped on the gasket.

On 1100, 1210, 1310, 1510, 1710, 1910 and 2110 models, head gaskets of different thicknesses are available for service and selection of correct gasket thickness is based on the distance pistons protrude above face of cylinder block at TDC. Use a dial indicator or other suitable means to measure height of each piston above cylinder block surface as shown in Fig. 50. Use the measurement from piston that has the highest protrusion and select appropriate thickness head gasket as indicated in chart shown in Fig. 51.

NOTE: On 1210 and 1310 models, the head gasket thickness is stamped on the top face of the gasket. On all other models, the last four digits of head gasket part number are stamped on top face of gasket.

Install head gasket with side marked TOP facing up. Lubricate threads of cylinder head retaining nuts and cap screws, then tighten in three steps to specified torque as listed below. Tighten cap screws in sequence shown in appropriate Fig. 52A through Fig. 52F.

Model	Specified Torque
1100	146-152 N·m
	(108-112 ft.-lbs.)
1200, 1300	150-155 N·m
	(111-114 ft.-lbs.)
1110	128-132 N·m
	(94 ft.-lbs.)
1210, 1310	48 N·m
	(35 ft.-lbs.)
1500, 1700	150-155 N·m
	(111-114 ft.-lbs.)

1510
10 mm Bolts . . . . .61 N·m
(45 ft.-lbs.)
12 mm Bolts . . . . .95 N·m
(70 ft.-lbs.)
1710
10 mm Bolts . . . . .61 N·m
(45 ft.-lbs.)
14 mm Bolts . . . . .129 N·m
(95 ft.-lbs.)
1900
Small Nuts (6 used) . . . . . . .60-65 N·m
(44-48 ft.-lbs.)
Large Nuts (11 used) . . .150-155 N·m
(111-114 ft.-lbs.)
1910, 2110 . . . . .95 N·m
(70 ft.-lbs.)

Complete installation, then adjust rocker arm to valve clearance as outlined in paragraph 28.

VALVE CLEARANCE

All Models

28. Clearance between rocker arm and end of valve stem should be checked and adjusted with engine cold and not running and with piston at TDC on compression stroke. Recommended clearance is 0.20 mm (0.008 inch) for intake and exhaust on 1110, 1210, 1310, 1510 and 1710 models. On all other models, recommended clearance is 0.30 mm (0.012 inch) for intake and exhaust.

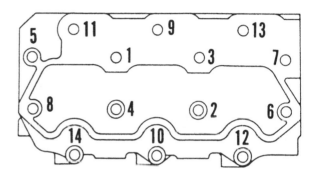

MODELS 1210,1310

Fig. 52B – On Models 1210 and 1310, tighten cylinder head bolts in sequence shown to specified torque.

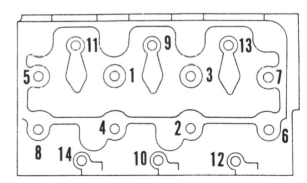

MODEL 1510

Fig. 52C – On Model 1510, tighten cylinder head bolts in sequence shown to specified torque.

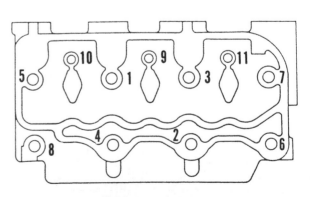

MODEL 1710

Fig. 52D – On Model 1710, tighten cylinder head bolts in sequence shown to specified torque.

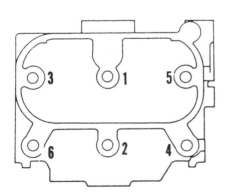

MODEL 1110

Fig. 52A – On Model 1110, tighten cylinder head bolts in sequence shown to specified torque.

ROCKER ARMS AND PUSH RODS

All Models

29. A single rocker arm shaft (10 – Fig. 47 and 48) is used to support all the rocker arms on 1210, 1310, 1510 and 1710 models. All other models use an individual support bracket and shaft (6 – Fig. 45, 46 or 49) to separately support rocker arms of each cylinder. Note that rocker arms and push rods should be reinstalled in their original location if being reused.

Inspect all parts for wear or damage. Clearance between rocker shaft and bore of rocker arm should not exceed 0.20 mm (0.008 inch). Measure shaft OD and rocker arm ID and renew parts that do not meet the following wear limits.

Model	Wear Limit
1100-1110-1200-1300-1500-1700-1900	
Shaft OD	13.55 mm
	(0.534 in.)
Rocker Arm ID	13.72 mm
	(0.540 in.)
1210-1310	
Shaft OD	11.57 mm
	(0.456 in.)
Rocker Arm ID	11.73 mm
	(0.462 in.)
1510-1710	
Shaft OD	17.55 mm
	(0.691 in.)
Rocker Arm ID	17.83 mm
	(0.702 in.)

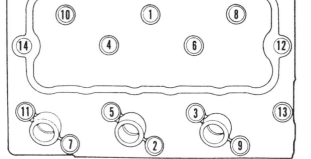

MODEL 1910

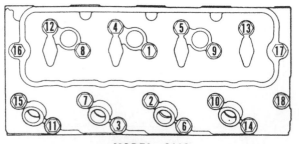

MODEL 2110

Fig. 54 — View of valve face and stem showing 45° face angle (A), chamfer (C) and margin (M). Refer to text for specifications.

1910-2110	
Shaft OD	13.55 mm
	(0.534 in.)
Rocker Arm ID	13.72 mm
	(0.540 in.)

Valve contacting surface of rocker arms can be resurfaced using an oil stone to remove slight wear. Check push rods for straightness by rolling on a flat surface. Inspect ends of push rods for excessive wear. Renew push rods as necessary.

VALVE SPRINGS

All Models

30. Valve springs are interchangeable for intake and exhaust valves. Renew springs if discolored, distorted or if they fail to meet the following specifications:

Models 1210-1310

Free Length (Min.)	33.5 mm
	(1.319 in.)
Out of Square (Max.)	1.8 mm
	(0.070 in.)

Fig. 52E — On Model 1910, tighten cylinder head bolts in sequence shown to specified torque.

Fig. 52F — On Model 2110, tighten cylinder head bolts in sequence shown to specified torque.

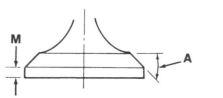

Test load	69-89 N
	(15.5-20 lbs.)
Test Length	30.4 mm
	(1.2 in.)

All Other Models

Free Length (Min.)	44 mm
	(1.732 in.)
Out of Square (Max.)	2 mm
	(0.080 in.)
Test Load	129-145 N
	(29-32.5 lbs.)
Test Length	38 mm
	(1.5 in.)

VALVES, GUIDES AND SEATS

All Models

31. Valve face angle (A – Fig. 54) is 45° for intake and exhaust. Valve should be renewed if valve head margin (M) is less than 0.05 mm (0.020 inch) on 1210 and 1310 models or 1.0 mm (0.040 inch) on all other models after refacing. Renew valve if margin is not equal all the way around the valve, which would indicate a bent valve. When refacing valve stem end, be sure the 0.80 mm (0.030 inch) chamfer (C) is maintained. Measure diameter of valve stem at top, center and bottom and renew if excessively worn.

Valve seat angle (A – Fig. 55) is 45° for intake and exhaust. After refacing seat, apply Prussian Blue to valve seat or mark valve face or seat with a soft

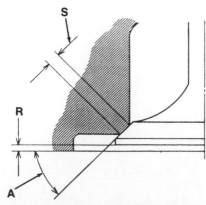

Fig. 55 — Drawing of valve and seat showing 45° seat angle (A), valve recession (R) and seat width (S). Refer to text for specifications.

lead pencil. Install valve and rotate valve against seat while applying light pressure against the valve. Seat contact point should be in center of valve face and seat width should be within specified limits. If necessary, use a 30° stone to lower seat contact point and use a 60° stone to raise seat contact point. Both stones will narrow the seat width. Check for correct recession (R) of valve head below machined surface of cylinder head. Renewable valve seat inserts are available on 1110, 1510, 1710, 1900, 1910 and 2110 models.

Measure valve stem diameter and valve guide bore to determine stem to guide clearance. Renewable valve guides are used on 1910 and 2110 models. On all other models, cylinder head must be renewed if guides are excessively worn. When installing new guides on 1910 and 2110 models, press guides in from the top until upper end of guide is flush with valve cover mounting surface of cylinder head as shown in Fig. 56.

Check valves and seats against the following specifications:

Valve Seat Width

Models 1100-1200-1300
    Standard..............1.2-1.5 mm
        (0.047-0.059 in.)
    Maximum ....2.5 mm
        (0.100 in.)
Models 1910-2110
    Standard..............1.7-2.5 mm
        (0067-0.098 in.)
    Maximum ....3.0 mm
        (0.120 in.)
All Other Models
    Standard..............1.2-1.5 mm
        (0.047-0.059 in.)
    Maximum ....2.0 mm
        (0.080 in.)

Stem Diameter-Intake

Models 1210-1310
    Standard ....6.955-6.97 mm
        (0.2738-0.2744 in.)
    Minimum................6.89 mm
        (0.271 in.)
All Other Models
    Standard..........7.955-7.97 mm
        (0.313-0.314 in.)
    Minimum................7.88 mm
        (0.310 in.)

Stem Diameter-Exhaust

Models 1210-1310
    Standard ....6.95-6.96 mm
        (0.2736-0.2740 in.)
    Minimum................6.84 mm
        (0.269 in.)
All Other Models
    Standard..........7.95-7.96 mm
        (0.313-0.314 in.)
    Minimum................7.85 mm
        (0.309 in.)

Stem to Guide Clearance-Intake

Models 1910-2110
    Standard...........0.06-0.085 mm
        (0.0025-0.0035 in.)
    Maximum ....0.20 mm
        (0.008 in.)
All Other Models
    Standard ....0.03-0.06 mm
        (0.001-0.0025 in.)
    Maximum ....0.20 mm
        (0.008 in.)

Stem to Guide Clearance-Exhaust

Models 1910-2110
    Standard ....0.06-0.85 mm
        (0.0025-0.0035 in.)
    Maximum ....0.25 mm
        (0.010 in.)
All Other Models
    Standard...........0.04-0.065 mm
        (0.0015-0.0025 in.)
    Maximum ....0.25 mm
        (0.010 in.)

Valve Recession From Face of Cylinder Head

Models 1100-1110-1200-1300
    Standard..............1.2-1.5 mm
        (0.047-0.059 in.)
    Maximum ....2.5 mm
        (0.100 in.)
Models 1210-1310
    Standard............0.85-1.15 mm
        (0.0335-0.0455 in.)
    Maximum ....2.15 mm
        (0.085 in.)
Models 1500-1700-1900
    Standard..............0.7-1.0 mm
        (0.028-0.039 in.)
    Maximum ....2.0 mm
        (0.080 in.)
Model 1510
    Standard..............0.8-1.1 mm
        (0031-0.043 in.)
    Maximum ....1.8 mm
        (0.070 in.)
Model 1710
    Standard.............1.05-1.35 mm
        (0.041-0.053 in.)
    Maximum ....2.0 mm
        (0.080 in.)
Model 1910
    Standard..............1.0-1.2 mm
        (0.039-0.047 in.)
    Maximum ....2.0 mm
        (0.080 in.)
Model 2110
    Standard.............0.07-1.0 mm
        (0.028-0.039 in.)
    Maximum ....2.0 mm
        (0.080 in.)

VALVE TIMING

All Models

32. The valve timing can be checked after removing timing gear case as outlined in paragraph 33 or 34. Refer to Fig. 57 for 1100, 1110, 1200 and 1300 models; Fig. 58 for 1210 and 1310 models; Fig. 59 for 1510 and 1710 models; Fig. 60 for 1500, 1700, 1900, 1910 and 2110 models. Camshaft is properly timed to crankshaft when timing marks (A and B) on crankshaft gear, idler gear and camshaft gear are aligned as shown.

Injection pump camshaft is keyed to drive gear (2—Fig. 57) on 1100, 1110, 1200 and 1300 models. The injection pump will be correctly timed to engine when timing gear marks are aligned as shown in Fig. 57.

The injection pump is activated by lobes on the engine camshaft on 1210 and 1310 models. Injection timing will be correct when timing gear marks are aligned as shown in Fig. 58.

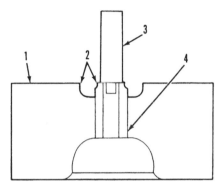

Fig. 56—When renewing valve guides on 1910 and 2110 models, press new guide in from the top until upper end of guide is flush with valve cover mounting surface of cylinder head.
1. Cylinder head   3. Piloted driver
2. Flush   4. Valve guide

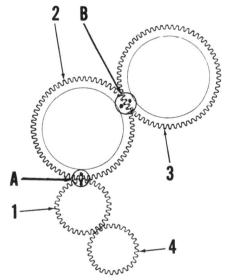

Fig. 57—Drawing of engine timing gears typical of 1100, 1110, 1200 and 1300 models. Valve timing is correct when timing marks (A and B) are aligned as shown.
1. Crankshaft gear   3. Camshaft gear
2. Injection pump gear   4. Oil pump gear

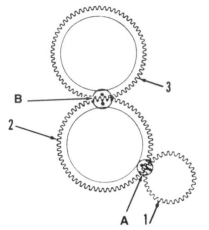

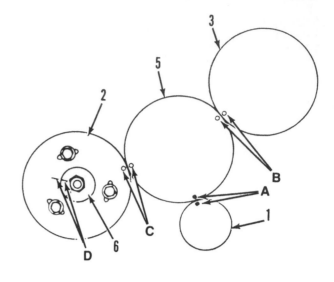

Fig. 59 — Drawing of timing gears typical of Model 1710. Model 1510 is similar. Valve timing is correct when timing marks (A and B) are aligned as shown. Injection pump timing is correct when timing marks (C and D) are aligned.

1. Crankshaft gear
2. Injection pump gear
3. Camshaft gear
5. Idler gear
6. Pump drive coupling

Fig. 58 — Drawing of engine timing gears typical of 1210 and 1310 models. Valve timing is correct when timing marks (A and B) are aligned as shown.

1. Crankshaft gear
2. Idler gear
3. Camshaft gear

On 1510 and 1710 models, injection pump drive coupling (6 – Fig. 59) is attached to pump drive gear. Injection pump should be correctly timed to engine when timing gear marks (A, B and C) are aligned and timing marks (D) on pump drive coupling (6) and gear (2) are aligned as shown.

On 1500, 1700, 1900, 1910 and 2110 models, injection pump drive shaft is attached to pump drive coupling (64 – Fig. 61) and coupling is bolted to drive gear (2). Injection pump drive gear timing should be checked after valve timing marks (A and B) on crankshaft gear, idler gear and camshaft gear are aligned. Rotate crankshaft counterclockwise until mark (A) on idler gear (which was aligned with mark on crankshaft gear) is now aligned with timing mark (C) on injection pump gear. Injection pump is properly timed to engine when timing gear marks (A and C) are aligned and marks (D) on pump drive coupling (64) and drive gear (2) are aligned as shown.

If injection pump, drive coupling or drive gears are renewed on 1500, 1510, 1700, 1710, 1900, 1910 or 2110 models, injection pump timing should be checked and adjusted as outlined in DIESEL FUEL SYSTEM section.

TIMING GEAR CASE

Models 1100-1110-1200-1300

33. To remove timing gear case (7 – Fig. 64), first drain cooling system and engine oil. Remove the radiator, coolant fan and fan pulley, alternator and water pump. Disconnect fuel lines from injection pump and immediately plug all openings in fuel system. Remove

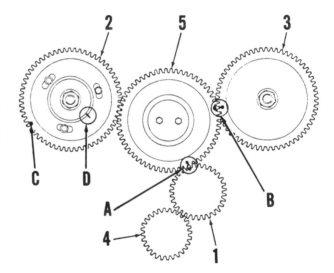

Fig. 60 — Drawing of timing gears typical of 1500, 1700, 1900, 1910 and 2110 models. Valve timing is correct when timing marks (A and B) are aligned as shown. Refer to Fig. 61 for timing of injection pump gear.

1. Crankshaft gear
2. Injection pump gear
3. Camshaft gear
4. Oil pump gear
5. Idler gear

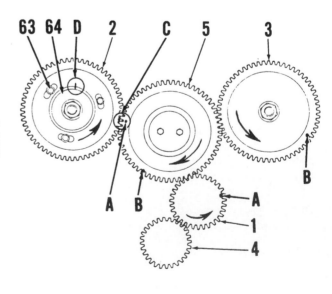

Fig. 61 — Drawing of timing gears positioned so injection pump drive gear timing marks (C and D) can be checked. Refer to text.

injection pump mounting cap screws, set pump control lever in middle of travel and remove injection pump. Be sure to retain any shims used between injection pump and timing gear case for use in installation. Remove hydraulic pump assembly. Remove crankshaft pulley. Remove camshaft gear cover (46 – Fig. 64), nut (45) and governor weight assembly (43). Remove injection pump camshaft bearing cover (51). It may be necessary to bump injection pump camshaft (49) back as gear case is pulled forward. Unbolt and remove timing gear case (7) from front of engine.

To install gear case, reverse the removal procedure. Note that crankshaft pulley is splined to crankshaft and timing marks on pulley and end of crankshaft must be aligned. Tighten pulley retaining cap screw to 48-58 N·m (36-43 ft.-lbs.) torque.

Models 1210-1310

34. To remove timing gear case, first drain cooling system and engine oil. Remove the radiator and cooling fan.

Remove the exhaust muffler and alternator. Remove the hydraulic pump. Remove injection pump high pressure fuel lines. Disconnect throttle control rod from governor lever. Remove injection pump mounting bolts, then raise injection pump and disconnect governor link from pump control rack. It is not necessary to remove injection pump. Remove the crankshaft pulley. Unbolt and remove timing gear case from front of engine.

To reinstall gear case, reverse the removal procedure. Be sure that dowel pin (2 – Fig. 67) in gear case engages hole in the oil pump front cover. Make certain that timing marks on crankshaft pulley and end of crankshaft are aligned.

Models 1500-1510-1700-1710-1900-1910-2110

35. To remove timing gear case (7 – Fig. 65), first drain cooling system and engine oil. Remove the radiator and cooling fan. Remove the alternator assembly. On 1510 and 1710 models, remove the water pump. On 2110

models, remove the power steering pump (if equipped). On all models, remove the hydraulic pump and filter assembly. Remove the crankshaft pulley. Unbolt and remove timing gear case from front of engine.

To reinstall, reverse the removal procedure. Be sure that timing marks on crankshaft pulley and end of crankshaft are aligned. Tighten pulley retaining cap screw to 49-59 N·m (36-43 ft.-lbs.) torque.

CAMSHAFT AND CAM FOLLOWERS

Models 1100-1110-1200-1300

37. To remove the camshaft, it is necessary to raise the mushroom type cam followers away from cam lobes so that camshaft can be withdrawn to the front. If proper tools are available to raise and hold the cam followers up, engine can remain attached to the clutch housing.

Drain engine coolant and oil. Refer to paragraph 26 and remove the engine.

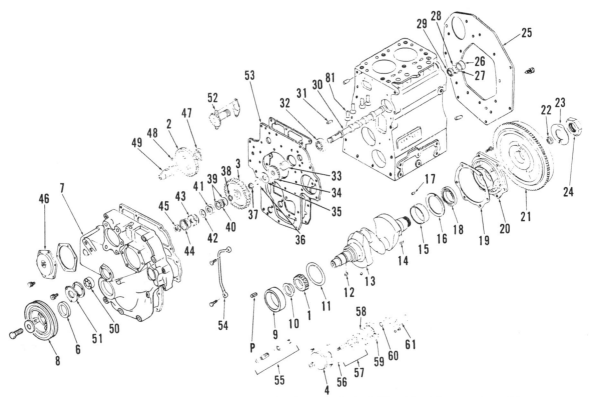

Fig. 64—Exploded view of timing gears and related parts typical of 1100, 1110, 1200 and 1300 models.

1. Crankshaft gear	12. Key	23. Lock plate
2. Injection pump cam gear	13. Crankshaft	24. Nut
3. Valve camshaft gear	14. Key	25. Rear plate
4. Oil pump drive gear	15. Rear main bearing	26. Plug
5. Crankshaft front seal	16. Rear thrust washer	27. Spacer
6. Crankshaft front seal	17. Pin	28. Snap ring
7. Timing gear housing	18. Rear oil seal	29. Camshaft rear bearing
8. Crankshaft pulley	19. Gasket	30. Camshaft
9. Front main bearing	20. Rear Housing	31. Key
10. Oil slinger	21. Flywheel	32. Front bearing
11. Front thrust washer	22. Pilot bearing	

33. Spacer	43. Weight assy.
34. Tachometer drive gear	44. Oil slinger
35. Washer	45. Nut
36. Shim	46. Cover
37. Snap ring	47. Ball bearing
38. Snap ring	48. Key
39. Bearing races	49. Injection pump cam
40. Needle thrust bearing	50. Ball bearing
41. Slider	51. Cover
42. Shim	52. Tachometer drive
	53. Front plate

54. Oil line
55. Oil pressure relief valve
56. Oil pump cover
57. Pump rotors & shaft
58. "O" ring
59. Housing
60. "O" ring
61. Inlet
81. Cam followers

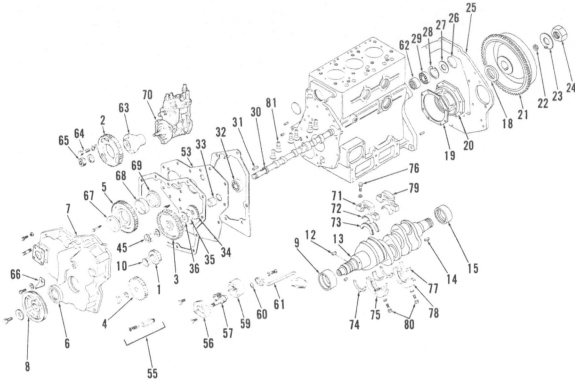

Fig. 65 — Exploded view of Model 1900 timing gears and related parts. Models 1500, 1510, 1700, 1710 and 1910 are similar.

1. Crankshaft gear
2. Injection pump cam gear
3. Valve camshaft gear
4. Oil pump drive gear
5. Idler gear
6. Crankshaft front oil seal
7. Timing gear housing
8. Crankshaft pulley
9. Front main bearing
10. Oil slinger
12. Key
13. Crankshaft
14. Key
15. Rear main bearing
18. Rear oil seal
19. Gasket
20. Rear housing
21. Flywheel
22. Pilot bearing
23. Lock plate
24. Nut
25. Rear plate
26. Plug
27. Spacer
28. Snap ring
29. Camshaft rear bearing
30. Camshaft
31. Key
32. Front bearing
33. Spacer
34. Tachometer drive gear
35. Washer
36. Shim
45. Nut
53. Front plate
55. Oil pressure relief valve
56. Oil pump cover
57. Oil pump shaft & gears
59. Housing
60. "O" ring
61. Inlet
62. Camshaft center bearing
63. Injection pump hub
64. Gear retaining screws
65. Hub retaining nut
66. Timing pointer
67. Plate
68. Idler gear bushing
69. Idler shaft
70. Injection pump
71. Top of second main bearing housing
72. Bearing insert
73. Top thrust bearing
74. Lower thrust bearing
75. Bottom of second main bearing housing
76. Main bearing housing clamp screws
77. Bearing insert
78. Bottom of third main bearing housing
79. Top of third main bearing housing
80. Main bearing housing retaining screws
81. Cam followers

Remove rocker arm cover, rocker arms and push rods. Remove timing gear case as outlined in paragraph 33. Remove timing gears, oil pump and front plate (53 – Fig. 64). Turn engine upside down or use suitable tools to hold cam followers away from camshaft, then withdraw camshaft from the front. Camshaft rear bearing (29) can be removed from cylinder block after removing clutch, flywheel, rear plate (25), plug (26) and snap ring (28).

Cam followers (81) can be removed from below after camshaft is removed.

Be sure to identify cam followers in order of removal so they can be reinstalled in their original locations if reused.

Inspect camshaft and cam followers for wear, scoring, chipping or other damage. Standard height of camshaft

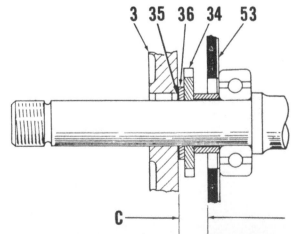

Fig. 66 — Cross section of camshaft and front bearing showing position of shims (36) which are used to adjust clearance (C) between drive gear (3) and front plate (53) on 1100, 1110, 1200 and 1300 models. Clearance should be 13.20-13.40 mm (0.6520-0.528 inch).

Fig. 67 — When installing timing gear case (1) on 1210 and 1310 models, be sure that dowel pin (2) in case engages hole in oil pump cover.

lobes is 36.02-36.07 mm (1.418-1.420 inches) and minimum allowable height is 35.65 mm (1.404 inches). Camshaft runout should not exceed 0.10 mm (0.004 inch). Cam followers should be renewed if camshaft is renewed.

When installing camshaft, measure distance (C – Fig. 66) between the front face of engine front plate (53) and rear face of camshaft gear (3). If clearance is nt 13.2-13.4 mm (0.520-0.528 inch), add or remove shims (36) as required before making final assembly.

Turn crankshaft and camshaft until injection pump gear can be installed with both sets of timing marks aligned as shown in Fig. 57.

Models 1210-1310

38. The camshaft can be removed without separating engine from transmission. However, if camshaft rear bearing (2 – Fig. 68) is to be renewed, engine must be removed from tractor and clutch, flywheel and rear plate removed.

To remove camshaft (3), remove radiator, timing gear case and fuel injection pump. Remove cylinder head, then lift barrel type cam followers out top of cylinder block. Be sure to identify position of each cam follower so they can be reinstalled in original positions if reused.

Withdraw camshaft with governor weight (10).

Remove collar (15), slider (14) and shim (13). Remove snap ring (12), then withdraw governor weight (10), camshaft gear (8) and tachometer gear (7) from camshaft.

Inspect camshaft and cam followers for wear, scoring or other damage. Standard height of camshaft valve lobes is 26.445-26.500 mm (1.041-1.043 inches) and minimum allowable height is 26.10 mm (1.027 inches). Standard height of camshaft fuel lobes is 33.94-34.06 mm (1.335-1.341 inches) and minimum allowable height is 33.80 mm (1.330 inches). Camshaft runout should not exceed 0.10 mm (0.004 inch). Cam followers should be renewed if camshaft is renewed.

When installing camshaft assembly, be sure to align timing gear marks as shown in Fig. 58.

Models 1510-1710

39. The camshaft can be removed without separating the engine from transmission. However, if camshaft rear bearing (4 – Fig. 69) is to be renewed, engine must be removed from tractor and clutch, flywheel and rear plate removed.

To remove camshaft, remove radiator, timing gear cover and cylinder head. Lift the barrel type cam followers (5) out from the top. Cam followers should be reinstalled in their original positions if reused. Remove hydraulic system pump,

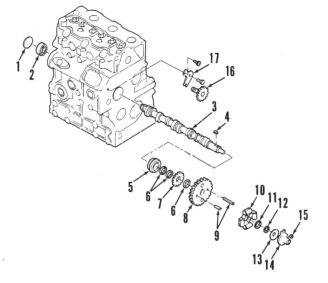

Fig. 68 – Exploded view of camshaft and related components used on 1210 and 1310 models.

1. Plug
2. Bearing
3. Camshaft
4. Key
5. Bearing
6. Spacers
7. Tachometer gear
8. Camshaft gear
9. Pins
10. Governor weight
11. Shim
12. Snap ring
13. Shim
14. Slider
15. Collar
16. Tachometer drive gear
17. Retainer plate

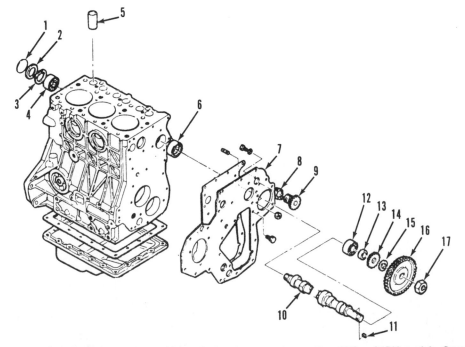

1. Explansion plug
2. Spacer
3. Snap ring
4. Rear bearing
5. Cam follower
6. Center bearing
7. Front plate
8. Gasket
9. Tachometer driven gear
10. Camshaft
11. Key
12. Front bearing
13. Collar
14. Tachometer drive gear
15. Spacer
16. Camshaft gear
17. Nut

Fig. 69 – Exploded view of camshaft and related components used on 1510 and 1710 models. Center bearing (6) is used on 1710 models.

injection pump and drive gear (1 – Fig. 70), idler gear (3) and shaft, camshaft gear (4), engine oil pump (5), oil pressure relief valve (6) and oil transfer tube (7). Remove engine front plate, then withdraw camshaft.

Inspect camshaft and cam followers for wear, scoring or other damage. Standard height of camshaft valve lobes is 36.02-36.07 mm (1.418-1.420 inches) and minimum allowable height is 35.5 mm (1.398 inches). Camshaft runout should not exceed 0.10 mm (0.004 inch). Cam followers should be renewed if camshaft is renewed.

To reinstall camshaft, reverse the removal procedure. Align timing gear marks as outined in paragraph 32.

Models 1500-1700-1900-1910-2110

40. To remove camshaft, it is necessary to move the mushroom type cam followers away from the cam lobes so camshaft can be withdrawn from the front. This is normally accomplished by turning the engine upside down to allow cam followers to fall away from the camshaft. However, if suitable tools are available to raise and hold the cam followers up, the camshaft can be removed without removing engine from the tractor.

Drain engine coolant and oil. Remove engine as outlined in paragraph 26. Remove rocker arm cover, rocker arms and supports, identifying the rocker assemblies so they can be reinstalled in their original position. Remove timing gear front cover, injection pump and drive gear, camshaft gear, engine oil pump and drive gear, oil transfer tube and tachometer driven gear. Remove engine front plate, then withdraw camshaft and front bearing from cylinder block. Remove clutch, flywheel and engine rear plate for access to camshaft

rear bearing. Cam followers and camshaft center needle bearing (1900, 1910 and 2110 models) are removed from below after removing engine oil pan and balancer assembly (2110 models).

Inspect camshaft and cam followers for wear, scoring or other damage and renew if necessary. Standard height of camshaft valve lobes on 1500, 1700 and 1900 models is 38.017-38.072 mm (1.497-1.500 inches) and minimum allowable height is 37.50 mm (1.476 inches). Standard height of camshaft valve lobes on 1910 and 2110 models is 38.06-38.07 mm (1.498-1.500 inches) and minimum allowable height is 37.65 mm (1.482 inches). Camshaft runout should not exceed 0.10 mm (0.004 inch) on all models. Cam followers should be renewed if camshaft is renewed.

Lubricate camshaft, bearings and cam followers prior to installation. Align timing gear marks as outlined in paragraph 32.

ROD AND PISTON UNITS

All Models

41. Connecting rod and piston units can be removed from above after removing cylinder head, engine oil pan and oil pump pickup tube. On 2110 models, engine balancer assembly and idler gear must also be removed. On all models, be sure that carbon deposits and wear ridge (if present) are removed from top of cylinder before pushing piston out. Remove connecting rod cap, then push piston and rod assembly out top of block.

Note that the piston pin is a transitional fit in piston. Heating the piston in hot water will facilitate removal and installation of piston pin. Be sure to keep piston, connecting rod and cap together and identified for each cylinder so they

can be reinstalled in their original locations.

Piston and rod cap must be assembled correctly to connecting rod for proper operation. Refer to appropriate Fig. 73 through 78 for correct assembly of specific models. Be sure that markings on rod and cap (A) are together and that connecting rod and piston are installed in the engine as shown. Tighten connecting rod cap retaining screws to the following torque:

Model	Torque
1100-1110-1200-1300	24-27 N·m (18-20 ft.-lbs.)
1210-1310	29-34 N·m (22-25 ft.-lbs.)
1500-1700	80-85 N·m (57-61 ft.-lbs.)
1510	24-27 N·m (18-20 ft.-lbs.)
1710-1900	44-49 N·m (32-36 ft.-lbs.)
1910-2110	78-83 N·m (58-62 ft.-lbs.)

PISTONS, RINGS AND CYLINDER

All Models

42. The aluminum cam ground pistons are fitted with two compression rings and one oil control ring on 1210 and 1310 models. All other models use three compression rings and one oil ring. Pistons and rings are available in standard size and 0.5 mm (0.020 inch) and 1.0 mm (0.0040 inch) oversizes for all models except 1910 and 2110. Pistons and rings are available in standard size only for 1910 and 2110 models.

Fig. 70 – View of engine timing gears on Model 1710.

1. Injection pump drive gear
2. Coupling
3. Idler gear
4. Camshaft gear
5. Oil pump
6. Oil pressure relief valve
7. Oil transfer tube

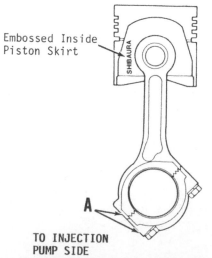

Embossed Inside Piston Skirt

TO INJECTION PUMP SIDE

Fig. 73 – View of piston and connecting rod correctly assembled for 1100, 1110, 1200 and 1300 models. Marks (A) on rod and cap must be aligned and face injection pump side of engine.

Check piston ring grooves for wear using new rings and a feeler gage to measure side clearance as shown in Fig. 80. Renew piston if clearance exceeds 0.25 mm (0.010 inch) for compression rings and 0.15 mm (0.006 inch) for oil control ring. Check pistons for wear, scratches or scoring and renew if necessary.

Check cylinder bore for wear, scoring or other damage. On all models except 1910 and 2110, cylinders can be rebored for installation of 0.5 mm (0.020 inch) or 1.0 mm (0.040 inch) oversize pistons.

On 1910 and 2110 models, renewable, wet type cylinder sleeves are used. The cylinder sleeves can be driven out top of cylinder block or removed using a suitable puller. Be sure to thoroughly clean cylinder block to remove rust and other deposits before installing new sleeve. Lubricate "O" ring seals, then carefully push sleeves into cylinder block until fully seated. Check protrusion of sleeve above face of cylinder block using a straightedge and feeler gage as shown in Fig. 81. Flange protrusion (A) should be 0.0-0.1 mm (0.0-0.004 inch) and lip protrusion (B) should be 0.90-1.05 mm (0.036-0.041 inch). If protrusion is excessive, check for foreign material in cylinder block counterbore. If sleeve flange dimension is correct, but lip protrusion is excessive, surface grind sleeve lip to obtain correct height.

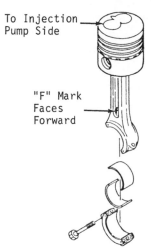

Fig. 78 — View of piston and connecting rod correctly assembled for 1910 and 2110 models. The "F" stamping on rod must face forward.

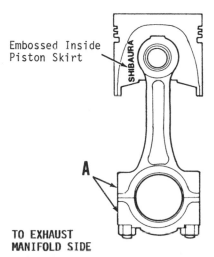

Fig. 74 — View of piston and connecting rod correctly assembled for 1210 and 1310 models. Marks (A) on rod and cap must be aligned and face away from injection pump side of engine.

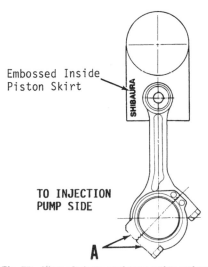

Fig. 76 — View of piston and connecting rod correctly assembled for 1510 and 1710 models. Marks (A) on rod and cap must be aligned and face injection pump side of engine.

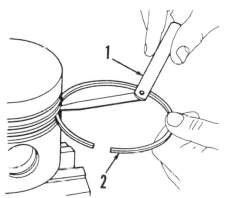

Fig. 80 — Check piston ring groove wear using a feeler gage (1) and new ring (2) as shown.

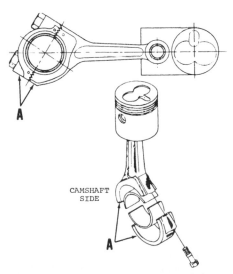

Fig. 75 — View of piston and connecting rod correctly assembled for 1500 and 1700 models. Marks (A) on rod and cap must be aligned and face camshaft side of engine.

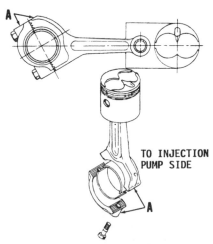

Fig. 77 — View of piston and connecting rod correctly assembled for 1900 models. Marks (A) on rod and cap must be aligned and face injection pump side of engine.

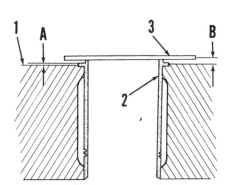

Fig. 81 — On 1910 and 2110 models, check cylinder sleeve (2) flange protrusion (A) and lip protrusion (B) above surface of cylinder block (1) using a straightedge (3) and feeler gage. Refer to text for specified dimensions.

Refer to the following specification data.

Models 1100-1110-1200
Piston Diameter*
  Standard...........74-89-74.92 mm
                    (2.948-2.950 in.)
  Minimum................74.70 mm
                    (2.941 in.)
Cylinder Bore
  Standard...........75.00-75.03 mm
                    (2.953-2.954 in.)
  Maximum ....75.20 mm
                    (2.961 in.)
Piston to Cylinder Clearance
  Desired ....0.08-0.13 mm
                    (0.003-0.005 in.)
Piston Pin Bore
  Standard........24.999-25.003 mm
                    (0.9842-0.9844 in.)
  Maximum ....25.016 mm
                    (0.9849 in.)
Piston Ring End Gap
  All Rings.............0.25-0.40 mm
                    (0.010-0.016 in.)

Model 1210
Piston Diameter*
  Standard.........71.91-71.94 mm
                    (2.831-2.832 in.)
  Minimum................71.70 mm
                    (2.823 in.)
Cylinder Bore
  Standard...........72.00-72.02 mm
                    (2.8346-2.8354 in.)
  Maximum ....72.20 mm
                    (2.8425 in.)
Piston to Cylinder Clearance
  Desired ....0.06-0.11 mm
                    (0.0024-0.0043 in.)
  Maximum ....0.25 mm
                    (0.010 in.)
Piston Pin Bore
  Standard........20.998-21.004 mm
                    (0.8267-0.8269 in.)
  Maximum ....21.016 mm
                    (0.827)
Piston Ring End Gap
  All Rings.............0.20-0.35 mm
                    (0.008-0.014 in.)

Model 1300
Piston Diameter*
  Standard...........79.87-79.90 mm
                    (3.1445-3.1455 in.)
  Minimum................79.70 mm
                    (3.138 in.)
Cylinder Bore
  Standard...........80.00-80.02 mm
                    (3.1495-3.1505 in.)
  Maximum ....81.20 mm
                    (3.197 in.)
Piston to Cylinder Clearance
  Desired ....0.10-0.15 mm
                    (0.004-0.006 in.)
  Maximum ....0.25 mm
                    (0.010 in.)

Piston Pin Bore
  Standard........24.999-25.003 mm
                    (0.9842-0.9844 in.)
Piston Ring End Gap
  All Rings.............0.20-0.45 mm
                    (0.008-0.018 in.)

Model 1310
Piston Diameter*
  Standard...........74.90-74.93 mm
                    (2.949-2.950 in.)
  Minimum................74.70 mm
                    (2.941 in.)
Cylinder Bore
  Standard...........75.00-75.02 mm
                    (2.953-2.954 in.)
  Maximum ....75.20 mm
                    (2.960 in.)
Piston to Cylinder Clearance
  Desired ....0.07-0.12 mm
                    (0.0028-0.0047 in.)
  Maximum ....0.25 mm
                    (0.010 in.)
Piston Pin Bore
  Standard........24.999-25.003 mm
                    (0.9842-0.9844 in.)
  Maximum ....25.016 mm
                    (0.985 in.)
Piston Ring End Gap
  All Rings.............0.25-0.40 mm
                    (0.010-0.016 in.)

Model 1500
Piston Diameter*
  Standard........84.855-84.877 mm
                    (3.3408-3.3416 in.)
  Minimum................84.70 mm
                    (3.335 in.)
Cylinder Bore
  Standard........85.000-85.02 mm
                    (3.3465-3.3472 in.)
  Maximum ....86.20 mm
                    (3.394 in.)
Piston to Cylinder Clearance
  Desired ....0.115-0.137 mm
                    (0.0045-0.0055 in.)
  Maximum ....0.30 mm
                    (0.012 in.)
Piston Pin Bore
  Standard........31.999-32.003 mm
                    (1.2598-1.2600 in.)
Piston Ring End Gap
  All Rings.............0.25-0.40 mm
                    (0.010-0.016 in.)

Model 1510
Piston Diameter*
  Standard........76.895-76.925 mm
                    (3.0275-3.0285 in.)
  Minimum................76.70 mm
                    (3.020 in.)
Cylinder Bore
  Standard...........77.00-77.02 mm
                    (3.0315-3.0325 in.)
  Maximum ....77.20 mm
                    (3.040 in.)

Piston to Cylinder Clearance
  Desired ....0.055-0.105 mm
                    (0.002-0.004 in.)
  Maximum ....0.25 mm
                    (0.010 in.)
Piston Pin Bore
  Standard........24.998-25.004 mm
                    (0.9842-0.9844 in.)
  Maximum ....25.02 mm
                    (0.985 in.)
Piston Ring End Gap
  Top Ring.............0.20-0.35 mm
                    (0.008-0.014 in.)
  All Other Rings ....0.15-0.30 mm
                    (0.006-0.012 in.)

Model 1700
Piston Diameter*
  Standard........89.845-89.875 mm
                    (3.5372-3.5384 in.)
  Minimum................89.70 mm
                    (3.532 in.)
Cylinder Bore
  Standard.........90.00-90.035 mm
                    (3.5435-3.5445 in.)
  Maximum ....91.20 mm
                    (3.590 in.)
Piston to Cylinder Clearance
  Desired ....0.125-0.190 mm
                    (0.005-0.007 in.)
  Maximum ....0.30 mm
                    (0.012 in.)
Piston Pin Bore
  Standard ....31.999-32.003 mm
                    (1.2598-1.260 in.)
Piston Ring End Gap
  All Rings.............0.20-0.45 mm
                    (0.008-0.018 in.)

Model 1710
Piston Diameter*
  Standard........83.865-83.895 mm
                    (3.3017-3.3029 in.)
  Minimum................83.70 mm
                    (3.295 in.)
Cylinder Bore
  Standard.........84.00-84.022 mm
                    (3.307-3.308 in.)
  Maximum ....84.20 mm
                    (3.315 in)
Piston to Cylinder Clearance
  Desired ....0.106-0.158 mm
                    (0.004-0.006 in.)
  Maximum ....0.30 mm
                    (0.012 in.)
Piston Pin Bore
  Standard........24.998-25.004 mm
                    (0.9842-0.9844 in.)
  Maximum ....25.02 mm
                    (0.985 in.)
Piston Ring End Gap
  All Rings.............0.25-0.40 mm
                    (0.010-0.016 in.)

Model 1900

Piston Diameter*
Standard........84.855-84.887 mm
(3.341-3.342 in.)
Minimum...............84.70 mm
(3.335 in.)

Cylinder Bore
Standard........85.00-85.035 mm
(3.3465-3.3478 in.)
Maximum ....86.2 mm
(3.394 in.)

Piston to Cylinder Clearance
Desired ....0.115-0.180 mm
(0.0045-0.0070 in.)
Maximum ....0.30 mm
(0.012 in.)

Piston Pin Bore
Standard ....24.999-25.003 mm
(0.9842-0.9844 in.)

Piston Ring End Gap
All Rings ....0.20-0.45 mm
(0.008-0.018 in.)

Models 1910-2110

Piston Diameter*
Standard..........84.88-84.91 mm
(3.342-3.343 in.)
Minimum...............84.72 mm
(3.335 in.)

Cylinder Bore
Standard.........85.00-85.022 mm
(3.3465-3.3475 in.)
Maximum ....85.18 mm
(3.354 in.)

Piston to Cylinder Clearance
Desired ....0.087-0.139 mm
(0.0035-0.0055 in.)
Maximum ....0.30 mm
(0.012 in.)

Piston Pin Bore
Standard ....32.0 mm
(1.260 in.)
Maximum ....32.08 mm
(1.263 in.)

Piston Ring End Gap
All Rings.............0.20-0.40 mm
(0.008-0.016 in.)

*Measure piston diameter at bottom of the skirt at right angle to piston pin bore.

Refer to appropriate Fig. 82, 83, 84 or 85 for installation of piston rings. If the expander for oil control ring uses a Teflon tube at end joint, tube and joint should be positioned at end gap of ring. If oil ring expander does not have Teflon tube, joint of expander (2) should be positioned opposite (180°) from ring gap. Top mark (1) on rings must be toward top of piston. Stagger ring end gaps approximately 90° from each other. Do not position a ring gap over piston pin bore. Lubricate cylinder bore and rings before installing piston and rod assemblies.

PISTON PIN

All Models

43. The full floating piston pin is retained in piston by snap rings. The pin is a transitional fit in piston bosses. Pin should have a 0.001 mm (0.00004 inch) interference fit to 0.007 mm (0.0003 inch) clearance in piston bosses. Maximum allowable clearance in piston bosses is 0.02 mm (0.0008 inch). Heating piston in hot water will facilitate removal and installation of piston pin.

Refer to the following specification data.

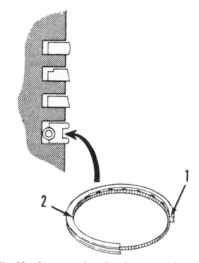

Fig. 82—Cross section showing correct installation of piston rings for 1100, 1110, 1200 and 1300 models.
1. Ring top mark (TP)
2. Joint of expander

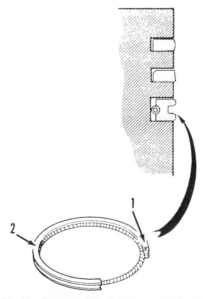

Fig. 83—Cross section showing correct installation of piston rings for 1210 and 1310 models.
1. Ring top mark (TP)
2. Joint of expander

Models 1100-1110-1200-1300-1310 -1900

Pin Diameter
Standard...........24.996-25.0 mm
(0.9841-0.9843 in.)
Wear Limit ....24.98 mm
(0.9834 in.)

Pin to Rod Bushing Clearance
Standard.............0.02-0.04 mm
(0.0008-0.0016 in.)
Wear Limit ....0.10 mm
(0.004 in.)

Model 1210

Pin Diameter
Standard...........20.996-21.0 mm
(0.8266-0.8268 in.)
Wear Limit ....20.98 mm
(0.826 in.)

Pin to Rod Bushing Clearance
Standard...........0.015-0.030 mm
(0.0006-0.0012 in.)
Wear Limit ....0.10 mm
(0.004 in.)

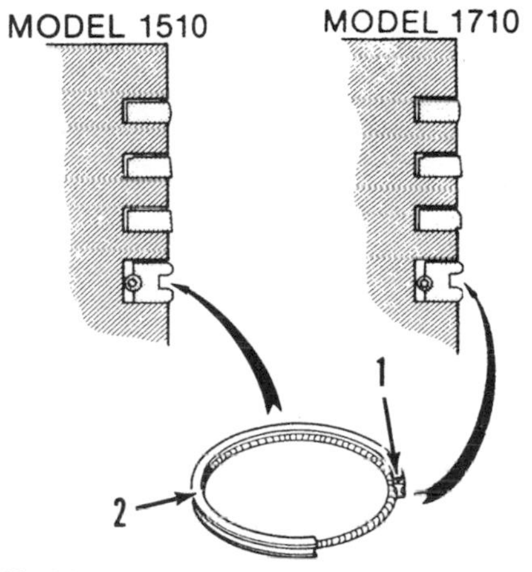

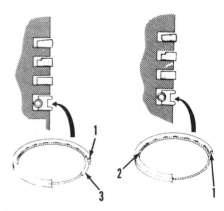

Fig. 84—Cross section showing correct installation of piston rings for 1510 and 1710 models.
1. Ring top mark (TP)
2. Joint of expander

Fig. 85—Cross section showing correct installation of piston rings for 1700, 1900, 1910 and 2110 models. If expander uses a Teflon tube (3), the tube and expander end joint should be positioned at end gap of oil ring.
1. Ring top mark (TP)
2. Joint of expander
3. Teflon tube

Models 1500-1700

Pin Diameter
Standard . . . . 31.996-32.0 mm
(1.2597-1.2598 in.)
Wear Limit . . . . 31.95 mm
(1.258 in.)
Pin to Rod Bushing Clearance
Standard . . . . 0.02-0.04 mm
(0.0008-0.0016 in.)
Wear Limit . . . . 0.10 mm
(0.004 in.)

Models 1510-1710

Pin Diameter
Standard . . . . 24.996-25.0 mm
(0.9841-0.9842 in.)
Wear Limit . . . . 24.98 mm
(0.9835 in.)
Pin to Rod Bushing Clearance
Standard . . . . 0.015-0.030 mm
(0.0006-0.0012 in.)
Wear Limit . . . . 0.08 mm
(0.003 in.)

Models 1910-2110

Pin Diameter
Standard . . . . 31.99-32.0 mm
(1.2595-1.2598 in.)
Wear Limit . . . . 31.97 mm
(1.2587 in.)
Pin to Rod Bushing Clearance
Standard . . . . 0.02-0.04 mm
(0.0008-0.0016 in.)
Wear Limit . . . . 0.15 mm
(0.006 in.)

Piston pin bushing in connecting rod is renewable. After new bushing is pressed into position, an oil hole must be drilled in top of bushing using the hole in the rod as a guide. Bushing must be reamed and finish honed to provide desired clearance for piston pin.

CONNECTING RODS AND BEARINGS

All Models

44. Connecting rod crankpin bearing inserts are available in standard size and 0.25 mm (0.010 inch) and 0.50 mm (0.020 inch) undersizes. Refer to paragraph 43 for installing and sizing piston pin bushings in connecting rod.

Refer to the following specification data.

Models 1100-1200-1300

Crankpin Diameter
Standard . . . . 47.951-47.970 mm
(1.8878-1.8886 in.)
Crankpin Bearing Diametral Clearance
Desired . . . . 0.035-0.096 mm
(0.0014-0.0037 in.)
Wear Limit . . . . 0.20 mm
(0.008 inch)

Model 1110

Crankpin Diameter
Standard . . . . 47.959-47.975 mm
(1.8882-1.8888 in.)
Crankpin Bearing Diametral Clearance
Desired . . . . 0.02-0.09 mm
(0.0008-0.0035 in.)
Wear Limit . . . . 0.2 mm
(0.008 in.)

Model 1210-1310

Crankpin Diameter
Standard . . . . 38.964-38.975 mm
(1.5340-1.5344 in.)
Crankpin Bearing Diametral Clearance
Desired . . . . 0.035-0.083 mm
(0.0014-0.0032 in.)
Wear Limit . . . . 0.2 mm
(0.008 in.)

Models 1500-1700

Crankpin Diameter
Standard . . . . 59.951-59.970 mm
(2.3603-2.3610 in.)
Crankpin Bearing Diametral Clearance
Desired . . . . 0.040-0.104 mm
(0.0016-0.0040 in.)
Wear Limit . . . . 0.20 mm
(0.008 in.)

Model 1510

Crankpin Diameter
Standard . . . . 47.964-47.975 mm
(1.8884-1.8888 in.)
Crankpin Bearing Diametral Clearance
Desired . . . . 0.035-0.085 mm
(0.0014-0.0033 in.)
Wear Limit . . . . 0.20 mm
(0.008 in.)

Model 1710

Crankpin Diameter
Standard . . . . 47.964-47.975 mm
(1.8884-1.8888 in.)
Crankpin Bearing Diametral Clearance
Desired . . . . 0.02-0.09 mm
(0.0008-0.0035 in.)
Wear Limit . . . . 0.20 mm
(0.008 in.)

Models 1910-2110

Crankpin Diameter
Standard . . . . 59.95-59.97 mm
(2.360-2.361 in.)
Crankpin Bearing Diametral Clearance
Desired . . . . 0.040-0.104 mm
(0.002-0.004 in.)
Wear Limit . . . . 0.20 mm
(0.008 in.)

Refer to paragraph 41 for assembly of piston to connecting rod. After installation, check connecting rod side play on crankshaft using a feeler gage. Normal side play is 0.10-0.30 mm (0.004-0.012 in.). Renew connecting rod if side play exceeds 0.70 mm (0.028 in.).

CRANKSHAFT AND MAIN BEARINGS

All Models

45. The crankshaft is supported at front and rear by sleeve type main bearings on all models except 1210 and 1310. On 1210 and 1310 models, the crankshaft is supported at the front by a sleeve type bearing and supported at the rear by a split type bearing insert and holder. On all three cylinder models, the crankshaft is also supported in the center by two split type bearing inserts and holders. On 2110 models, crankshaft is supported in the center by three split type bearing inserts and holders.

On 1100, 1110, 1200, 1300, 1500 and 1700 models, crankshaft end play is limited by thrust washers located at inner faces of front and rear main bearings. On 1210 and 1310 models, crankshaft end play is limited by thrust washers located at front and rear face of rear main bearing holder. On 1510 and 1710 models, crankshaft end play is limited by thrust washers located at front and rear of third (from front) main bearing holder. On 1900 and 1910 models, crankshaft end play is limited by thrust washers located at front and rear of second (from front) main bearing holder. On 2110 models, crankshaft end play is limited by thrust washers located at the center (third from front) main bearing holder.

To remove crankshaft and bearings, the engine must be removed from the tractor as outlined in paragraph 26. Remove cylinder head, oil pan, engine balancer (Model 2110) and piston and rod assemblies. Remove timing gear case, clutch, flywheel, engine rear plate and rear main bearing holder. On three and four cylinder models, remove cap screws retaining center main bearing holders. On all models, withdraw crankshaft from the rear.

Measure crankshaft journals for size, taper and out-of-round. Crankshaft should be reground to appropriate undersize or renewed if taper or out-of-round exceeds 0.05 mm (0.002 inch). Main bearings and connecting rod bearings are available in undersizes of 0.25 mm (0.010 inch) and 0.50 mm (0.020 inch) as well as standard size. Check crankshaft runout and straighten or renew crankshaft if runout exceeds 0.05 mm (0.002 inch). Refer to the following crankshaft specifications.

Models 1100-1200-1300

Main Journal Diameter
Standard . . . . 67.951-67.970 mm
(2.6752-2.6760 in.)

Main Bearing Clearance
Desired . . . . .0.040-0.118 mm
(0.0016-0.0046 in.)
Wear Limit . . . . .0.25 mm
(0.010 in.)
Crankpin Journal Diameter
Standard 47.951-47.970
(1.8878-1.8886 in.)
Crankpin Bearing Clearance
Desired . . . . .0.035-0.096 mm
(0.0014-0.0038 in.)
Wear Limit . . . . .0.20 mm
(0.008 in.)
Taper and Out-of-Round, All Journals
Wear Limit . . . . .0.05 mm
(0.002 in.)
End Play
Desired . . . . .0.10-0.45 mm
(0.004-0.018 in.)
Wear Limit . . . . .0.70 mm
(0.028 in.)

Model 1110
Main Journal Diameter
Standard . . . . .67.951-67.970 mm
(2.6752-2.6760 in.)
Main Bearings Clearance
Desired . . . . .0.04-0.10 mm
(0.0016-0.0039 in.)
Wear Limit . . . . .0.20 mm
(0.008 in.)
Crankpin Journal Diameter
Standard . . . . .47.959-47.975 mm
(1.8882-1.8888 in.)
Crankpin Bearing Clearance
Desired . . . . .0.02-0.09 mm
(0.0008-0.0035 in.)
Wear Limit . . . . .0.20 mm
(0.008 in.)
Taper and Out-of-Round,
All Journals
Wear Limit . . . . .0.025 mm
(0.001 in.)
End Play
Desired . . . . .0.10-0.40 mm
(0.004-0.016 in.)
Wear Limit . . . . .0.70 mm
(0.028 in.)

Models 1210-1310
Main Journal Diameter
Standard . . . . .45.964-45.975 mm
(1.8096-1.8100 in.)
Main Journal to Front
Bearing Clearance
Desired . . . . .0.039-0.106 mm
(0.0015-0.0040 in.)
Wear Limit . . . . .0.15 mm
(0.006 in.)
Main Journal to Center and
Rear Bearing Clearance
Desired . . . . .0.039-0.092 mm
(0.0015-0.0036 in.)
Wear Limit . . . . .0.20 mm
(0.008 in.)
Crankpin Journal Diameter
Standard . . . . .38.964-38.975 mm
(1.5340-1.5344 in.)

Crankpin Bearing Clearance
Desired . . . . .0.035-0.083 mm
(0.0014-0.0032 in.)
Wear Limit . . . . .0.20 mm
(0.008 in.)
Taper and Out-of-Round,
All Journals
Wear Limit . . . . .0.05 mm
(0.002 in.)
End Play
Desired . . . . .0.05-0.30 mm
(0.002-0.012 in.)
Wear Limit . . . . .0.70 mm
(0.028 in.)

Models 1500-1700
Main Journal Diameter
Standard . . . . .67.951-67.970 mm
(2.6752-2.6760 in.)
Main Bearing Clearance
Desired . . . . .0.056-0.134 mm
(0.0022-0.0052 in.)
Wear Limit . . . . .0.25 mm
(0.010 in.)
Crankpin Journal Diameter
Standard . . . . .59.951-59.970 mm
(2.3603-2.3610 in.)
Crankpin Bearing Clearance
Desired . . . . .0.040-0.104 mm
(0.0016-0.0040 in.)
Wear Limit . . . . .0.20 mm
(0.008 in.)
Taper and Out-of-Round,
All Journals
Wear Limit . . . . .0.05 mm
(0.002 in.)
End Play
Desired . . . . .0.10-0.45 mm
(0.004-0.018 in.)
Wear Limit . . . . .0.70 mm
(0.028 in.)

Models 1510-1710
Main Journal Diameter
Standard . . . . .67.957-67.970 mm
(2.6755-2.6760 in.)
Main Journal to Front and
Rear Bearing Clearance
Desired-1510 . . . . . . .0.050-0.116 mm
(0.0020-0.0045 in.)
1710 . . . . .0.04-0.10 mm
(0.0016-0.0039 in.)
Wear Limit –
1510 and 1710 . . . . .0.20 mm
(0.008 in.)
Main Journal to Center
Bearing Clearance
Desired – 1510 . . . . .0.060-0.112 mm
(0.0024-0.0044 in.)
1710 . . . . .0.070-0.134 mm
(0.0028-0.0053 in.)
Wear Limit –
1510 and 1710. . . . . .0.20 mm
(0.008 in.)
Crankpin Journal Diameter
Standard . . . . .47.964-47.975 mm
(1.8884-1.8888 in.)

Crankpin Bearing Clearance
Desired – 1510 . . . . .0.035-0.085 mm
(0.0014-0.0033 in.)
1710 . . . . .0.02-0.09 mm
(0.0008-0.0035 in.)
Wear Limit –
1510 and 1710 . . . . .0.20 mm
(0.008 in.)
Taper and Out-of-Round,
All Journals
Wear Limit . . . . 0.05 mm
(0.002 in.)
End Play
Desired . . . . .0.10-0.45 mm
(0.004-0.018 in.)
Wear Limit . . . . .0.70 mm
(0.028 in.)

Model 1900
Main Journal Diameter
Standard . . . . .67.951-67.970 mm
(2.6752-2.6760 in.)
Main Journal to Front and
Rear Bearing Clearance
Desired . . . . .0.056-0.134 mm
(0.0022-0.0052 in.)
Wear Limit . . . . .0.25 mm
(0.010 in.)
Main Journal to Center
Bearing Clearance
Desired . . . . .0.070-0.134 mm
(0.0028-0.0053 in.)
Wear Limit . . . . .0.25 mm
(0.010 in.)
Crankpin Journal Diameter
Standard . . . . .51.951-51.970 mm
(2.0453-2.0460 in.)
Crankpin Bearing Clearance
Desired . . . . .0.040-0.104 mm
(0.0016-0.0040 in.)
Wear Limit . . . . .0.20 mm
(0.008 in.)
Taper and Out-of-Round,
All Journals
Wear Limit . . . . .0.05 mm
(0.002 in.)
End Play
Desired . . . . .0.10-0.45 mm
(0.004-0.018 in.)
Wear Limit . . . . .0.70 mm
(0.028 in.)

Models 1910-2110
Main Journal Diameter
Standard . . . . .67.95-67.97 mm
(2.675-2.676 in.)
Main Journal to Front and
Rear Bearing Clearance
Desired . . . . .0.056-0.131 mm
(0.0022-0.0052 in.)
Wear Limit . . . . .0.20 mm
(0.008 in.)
Main Journal to Center
Bearing Clearance
Desired . . . . .0.070-0.134 mm
(0.0028-0.0052 in.)
Wear Limit . . . . .0.20 mm
(0.008 in.)

Crankpin Journal Diameter
 Standard . . . . 59.95-59.97 mm
  (2.3602-2.3610 in.)
Crankpin Bearing Clearance
 Desired . . . . 0.040-0.104 mm
  (0.0016-0.0040 in.)
 Wear Limit . . . . 0.20 mm
  (0.008 in.)
Taper and Out-of-Round,
All Journals
 Wear Limit . . . . 0.05 mm
  (0.002 in.)
End Play
 Desired . . . . 0.10-0.45 mm
  (0.004-0.018 in.)
 Wear Limit . . . . 0.70 mm
  (0.028 in.)

When renewing front and rear main bearings, be sure that oil hole in bearing is aligned with oil hole in cylinder block and rear cover as shown in Fig. 86 through 89. Notch (3) in bearings should be located on top. Press bearings into cylinder block or rear cover to proper depth using a suitable driver. Check bearing inside diameter after installation to make sure that bearing was not deformed and that journal to bearing clearance is within specifications. On three and four cylinder models, center main bearing clearance should be checked using Plastigage.

The cylinder block of early 1100 and 1300 models uses one pin to hold front thrust washer in position while late production 1100 and 1300 models use two pins. The early cylinder blocks can be modified to accommodate two pins if desired. To modify early cylinder block, drill two 4 mm (5/32 inch) diameter holes as indicated in Fig. 90. Do not attempt to use existing hole (B), but drill two new holes on 45° angle as indicted at (A) in the drawing. Holes (A) should be 42 mm (1.653 inches) from center of bearing bore as indicated at (D). The spring pins (P – Fig. 91) should be installed 1 mm (0.040 inch) below thrust surface of washer (11).

On all two cylinder models, install thrust washer in front of block. Lubricate main journals, then slide crankshaft into front main bearing. Install rear cover with main bearing, thrust washer and new oil seal. Tighten cover retaining cap screws to 46-54 N·m (35-40 ft.-lbs.) torque. Be sure that crankshaft is free to turn without binding. Measure crankshaft end play using a dial indicator. If end play exceeds 0.70 mm (0.028 inch), renew thrust washers.

On all three cylinder and four cylinder models, lubricate and assemble center main bearings, thrust washers and bearing holders on crankshaft. Be sure that chamfered side (5 – Fig. 92, 93, 94 or 95) of bearing holders (1) face toward front of crankshaft. Note that second (from

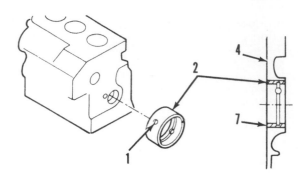

Fig. 86 — View showing correct installation of front main bearing on 1210 and 1310 models. Be sure that oil holes in bearing and cylinder block are aligned.
1. Oil hole
2. Front main bearing
4. Front face of block
7. Flush

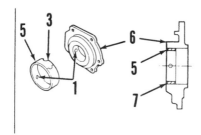

Fig. 87 — View showing correct installation of front and rear main bearings on 1510 models. Notch (3) in bearings goes to the top.

1. Oil hole
2. Front main bearing
3. Notch

4. Front face of block
5. Rear main bearing

6. Rear cover
7. Flush

Fig. 88 — View showing correct installation of front and rear main bearings on 1710 models. Notch (3) in bearings goes to the top.

A. 1.0 mm (0.040 in.)
1. Oil hole

2. Front main bearing
3. Notch

4. Front face of block
5. Rear main bearing

6. Rear cover
7. Flush

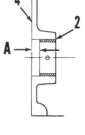
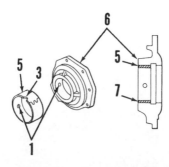

Fig. 89 — View showing correct installation of front and rear main bearings on 1910 and 2110 models. Notch (3) in bearings goes to the top.

A. 12.0 mm (0.472 in.)
1. Oil holes

2. Front main bearing
3. Notch

4. Front face of block
5. Rear main bearing

6. Rear cover
7. Flush

front) holder on 1210 and 1310 models has an identification mark (3—Fig. 92) on it to ensure correct assembly. Tighten bearing holder cap screws to a torque of 25-29 N·m (18-22 ft.-lbs.) on 1210 and 1310 models; 48-53 N·m (36-39 ft.-lbs.) on 1510 and 1710 models; 70-80 N·m (51-59 ft.-lbs.) on 1900, 1910 and 2110 models. Insert crankshaft assembly into cylinder block and front main bearing. Install bearing holder retaining bolts and tighten to a torque of 25-29 N·m (18-22 ft.-lbs.) on 1210 and 1310 models and 70-80 N·m (51-59 ft.-lbs.) on all other models.

Measure crankshaft end play using a dial indicator. If end play exceeds 0.70 mm (0.028 inch), renew thrust washers. Install oil seal and rear plate on 1210 and 1310 models and tighten retaining cap screws to 27-33 N·m (20-24 ft.-lbs.) torque. On all other models, install rear cover with main bearing and new oil seal. Tighten cover retaining cap screws evenly to 46-54 N·m (34-40 ft.-lbs.) torque.

On all models, complete installation by reversing removal procedure. Be sure timing gear marks are correctly aligned as outlined in paragraph 32. On 2110 models, refer to paragraph 47 for timing and installation of engine balancer. On all models, be sure timing marks on crankshaft pulley and end of crankshaft are aligned.

FLYWHEEL

All Models

46. To remove flywheel, first split tractor between engine and transmission as outlined in paragraph 101 and remove clutch assembly. On 1210, 1310 and 1510 models, remove cap screws and retainer plate attaching flywheel to crankshaft, then remove the flywheel. On all other models, loosen the flywheel retaining nut, but do not remove the nut from the crankshaft. While prying outward on flywheel, tap end of crankshaft with a brass drift and hammer to loosen flywheel from the tapered end of crankshaft. Remove retaining nut and flywheel.

Inspect flywheel and ring gear for excessive wear or other damage and renew if necessary. When installing a new ring gear, heat the gear evenly to a temperature of 120°-150° C (245°-300° F). Install the heated gear quickly making sure it is seated against shoulder of flywheel.

Install flywheel on crankshaft and tighten retaining cap screws to 56-69 N·m (41-51 ft.-lbs.) torque on 1210, 1310 and 1510 models. On all other models, tighten retaining nut to 343-441 N·m (253-325 ft.-lbs.) torque.

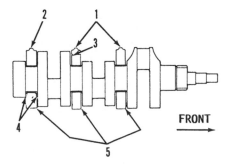

Fig. 92—On 1210 and 1310 models, thrust washers (4) are located on rear main bearing holder (2). Second bearing holder has an identification mark (3). Chamfered side (5) of bearing holders should be towards the front.

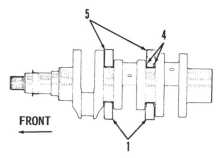

Fig. 93—On 1510 and 1710 models, thrust washers (4) are located on second (from front) main bearing holder. Chamfered side (5) of bearing holders should face forward.

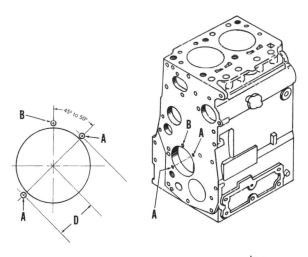

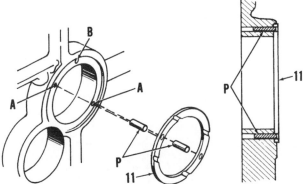

Fig. 90—Refer to text for modifying early 1100 and 1300 cylinder blocks for installation of two pins to retain front thrust washer. Refer also to Fig. 91.

A. New holes
B. Old hole
C. 42 mm (1.653 in.)

Fig. 91—Drawing at left side is inside cylinder block showing installation of front thrust washer (11) and pins (P). Pins should be 1.0 mm (0.040 inch) below thrust surface of washer as shown in cross section at right.

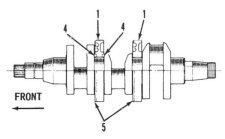

Fig. 94—On 1900 and 1910 models, thrust washers (4) are located on first (from front) main bearing holder. Chamfered side (5) of holders should be toward the front.

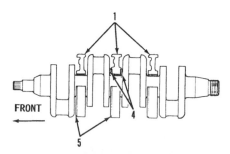

Fig. 95—On 2110 models, thrust washers (4) are located on center main bearing holder. Chamfered side (5) of bearing holders should be toward the front.

ENGINE BALANCER

Model 2110

47. To remove balancer assembly, first drain engine oil. If equipped with front wheel drive, remove front axle and drive shaft. Remove engine oil pan. Remove balancer bottom cover (9–Fig. 96). Remove balancer mounting cap screws, then lower balancer assembly from engine. Note that hollow dowels (6) are used to position balancer on engine block and may restrict removal of balancer. Retain shims (7), if used, for use in installation.

Remove snap ring (5–Fig. 97), thrust washer (4) and idler gear and bearing assembly (3). If renewal of idler gear shaft (2) is necessary, the engine must first be separated from transmission as outlined in paragraph 101. Remove clutch, flywheel, engine rear plate and rear main bearing cover. Push idler shaft rearward from engine block. A removal tool can be fabricated using a suitable size piece of pipe, a bolt, washer, nut and the dimensions shown in Fig. 98. Place guide tube (4–Fig. 99) over idler shaft (2) and position end of bolt (1) against block webbing, then turn the nut to force idler shaft out rear of block.

To disassemble balancer, remove upper cover (8–Fig. 96) and end cover (16). It is recommended that balancer gear backlash be checked prior to removal of gears (14). If backlash exceeds 0.15 mm (0.006 inch), gears should be renewed. Remove shaft retaining cap screws (1) and counterweight set screws (11). Mark the balancer gear shafts so they can be reinstalled in their original location if reused. Tap the shafts out the gear end of frame and withdraw counterweights (10). Remove bearings (3 and 4) from housing bores.

Inspect all parts for wear or damage and renew if necessary.

Fig. 97—View of balancer idler gear assembly.

1. Drive gear
2. Idler shaft
3. Idler gear & bearings
4. Thrust washer
5. Snap ring

To reassemble, reverse the disassembly procedure while noting the following special instructions. When installing balancer frame bearings (4–Fig. 100), be sure that shield side (S) of inner bearings face outward. Lubricate thrust washer wear surfaces with molybdenum disulfide grease. Install balancer shaft and gear assemblies and counterweights, aligning matching marks on gears for proper timing. Tighten counterweight set screws to 24-29 N·m (17-22 ft.-lbs.) torque. Be sure shafts turn freely.

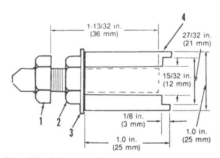

Fig. 98—Idler shaft removal tool can be fabricated using dimensions shown.

1. 11 mm bolt
2. Nut
3. Flat washer
4. Steel pipe

If idler gear shaft was removed, drive new shaft into cylinder block from the rear until rear of shaft is flush with rear counterbore. Install a new expansion plug in rear counterbore. Assemble idler gear, thrust washer and snap ring on idler shaft.

To install balancer assembly, first rotate crankshaft until No. 1 and 4

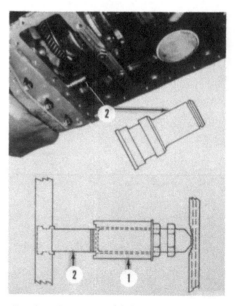

Fig. 99—Removal tool (1) is used to push idler shaft (2) rearward from cylinder block. Refer to text.

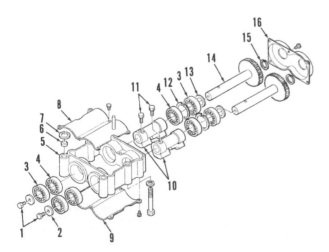

Fig. 96—Exploded view of engine balancer assembly used on Model 2110.

1. Retaining screws
2. Washers
3. Outer bearings
4. Inner bearings
5. Frame
6. Dowel guide
7. Shim
8. Upper cover
9. Lower cover
10. Counterweights
11. Set screws
12. Snap ring
13. Thrust washer
14. Shaft & gear
15. Snap ring
16. End cover

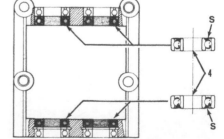

Fig. 100—When installing inner bearings (4) in balancer frame, be sure that sealed side (S) of bearings face outward.

pistons are at top dead center. With balancer counterweights in a straight down position, install balancer assembly with original shims (if used) on engine block. Tighten retaining cap screws to 70-79 N·m (51-59 ft.-lbs.) torque. Install lower cover and oil pan.

OIL PUMP AND RELIEF VALVE

Models 1100-1110-1200-1300

48. The rotor type oil pump is located in a bore in the front of the cylinder block. The pump is driven by the camshaft gear.

To remove the oil pump, first remove the timing gear case as outlined in paragraph 33. Remove oil pump drive gear and oil pump mounting cap screws. Withdraw pump and oil pickup tube as an assembly.

Unbolt and remove pump front cover (4–Fig. 101). Withdraw inner and outer rotor (5 and 6). Check all parts for wear or damage. Use a feeler gage to check rotor clearances as shown in Figs. 102, 103 and 104. Refer to the following specifications and renew pump if necessary.

Rotor to Cover End Clearance
  Standard.............0.03-0.11 mm
                (0.001-0.004 in.)
  Wear Limit ....0.20 mm
                (0.008 in.)
Outer Rotor to Body Clearance
  Standard.............0.14-0.22 mm
                (0.006-0.009 in.)
  Wear Limit ....0.30 mm
                (0.012 in.)
Rotor to Rotor Clearance
  Standard.............0.01-0.15 mm
                (0.0004-0.006 in.)
  Wear Limit ....0.25 mm
                (0.010 in.)

Engine oil pressure relief valve (2) is serviced as an assembly and is not adjustable. Relief valve opening pressure should be 241-393 kPa (35-57 psi) at 2600 engine rpm with oil temperature at 80° C (175° F). Minimum oil pressure at idle speed is 117 kPa (17 psi).

Models 1210-1310

49. The rotor type oil pump is located inside the engine idler gear (11–Fig. 105). The idler gear is pinned to the pump outer rotor (8) and is driven by the crankshaft gear.

To remove oil pump, first remove timing gear case as outlined in paragraph 34. Remove "E" ring (1), then slide idler gear with pump components as an assembly off the pump shaft (4).

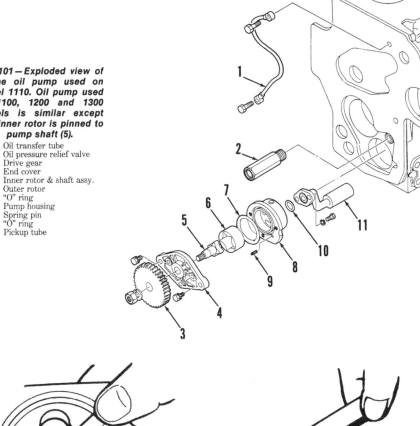

Fig. 101—Exploded view of engine oil pump used on Model 1110. Oil pump used on 1100, 1200 and 1300 models is similar except that inner rotor is pinned to pump shaft (5).

1. Oil transfer tube
2. Oil pressure relief valve
3. Drive gear
4. End cover
5. Inner rotor & shaft assy.
6. Outer rotor
7. "O" ring
8. Pump housing
9. Spring pin
10. "O" ring
11. Pickup tube

Fig. 102—Use a straightedge and feeler gage to check rotor to cover end clearance. Refer to text.

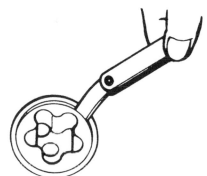

Fig. 103—Measure outer rotor to housing clearance using a feeler gage as shown. Refer to text.

The pump shaft and oil port block assembly (Fig. 106) is a press fit in bore of cylinder block. If renewal is required, special tool No. 11097 and a slide hammer may be used to pull port block from cylinder block. Special installing tool No. 11063 is available for proper installation of port block. The port block is in correct installed position when installing tool bottoms against engine block.

Inspect pump cover (6–Fig. 105), rotors (7 and 8) and port head for wear, scratches or scoring and renew if necessary. Use a feeler gage to check rotor to rotor clearance as shown in Fig. 104. Standard clearance is 0.01-0.15 mm (0.0004-0.006 in.) Renew rotors if clearance exceeds 0.25 mm (0.010 inch).

Engine oil pressure relief valve is located in right-hand side of cylinder block below the fuel injection pump. The relief valve is serviced as an assembly and is not adjustable. Relief valve open-

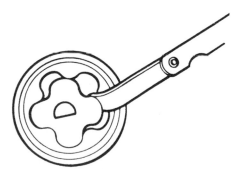

Fig. 104—Check clearance between inner rotor and outer rotor vanes using a feeler gage. Refer to text.

ing pressure should be 297-490 kPa (43-71 psi) at 2600 engine rpm with oil temperature at 80° C (175° F). Minimum oil pressure at idle speed is 48 kPa (7 psi).

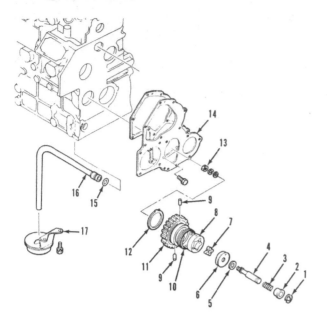

Fig. 105 — Exploded view of engine oil pump assembly used on 1210 and 1310 models.

1. Snap ring
2. Collar
3. Spring
4. Shaft
5. Shim
6. Cover
7. Inner rotor
8. Outer rotor
9. Pins
10. Spring
11. Idler gear
12. Thrust washer
13. Nut
14. Front plate
15. "O" ring
16. Pickup tube
17. Strainer

The engine oil pressure relief valve (2 – Fig. 107) is serviced as an assembly and is not adjustable. Relief valve opening pressure should be 241-393 kPa (35-57 psi) at 2600 engine rpm with oil temperature at 80° C (175° F).

Models 1500-1700-1900-1910-2110

51. The gear type engine oil pump is located in a bore in the front of engine block. The pump drive gear is driven by the crankshaft gear.

The timing gear case must be removed as outlined in paragraph 35 for access to pump gears. Remove drive gear (3 – Fig. 108) and front cover (4), then withdraw pump gears (5). Pump housing (6) can be removed from the front after engine oil pan and oil pickup tube are first removed

Inspect gears, cover and pump housing for wear, scratches and scoring and renew if necessary. Use a feeler gage to check pump gear clearances as shown in Figs. 109 and 110. Renew pump if pump gear to cover end clearance exceeds 0.10 mm (0.004 inch) or if gear to housing side clearance exceeds 0.15 mm (0.006 inch).

The engine oil pressure relief valve (2 – Fig. 108) is serviced as an assembly and is not adjustable. Relief valve opening pressure should be 241-393 kPa (35-57 psi) at 2600 engine rpm with oil temperature at 80° C (175° F). Minimum oil pressure at idle speed is 117 kPa (17 psi).

Fig. 106 — Oil pump shaft (2) and port block (1) assembly is a press fit in cylinder block on 1210 and 1310 models.

Rotor to Cover End Clearance
    Standard . . . . .0.03-0.11 mm
    (0.001-0.004 in.)
    Wear Limit . . . . .0.20 mm
    (0.008 in.)
Outer Rotor to Housing Clearance
    Standard . . . . .0.14-0.22 mm
    (0.006-0.009 in.)
    Wear Limit . . . . .0.30 mm
    (0.012 in.)
Rotor to Rotor Clearance
    Standard . . . . .0.01-0.15 mm
    (0.0004-0.006 in.)
    Wear Limit . . . . .0.25 mm
    (0.010 in.)

Models 1510-1710

50. The rotor type engine oil pump is located in a bore in front of engine block. The oil pump drive gear is driven by the camshaft gear. Timing gear case must be removed as outlined in paragraph 35 for access to the oil pump.

Remove oil pump drive gear. Remove oil pump mounting cap screws, then withdraw the oil pump assembly.

Remove front cover (4 – Fig. 107), drive shaft (5) and rotor assembly (6 and 8) from pump housing (11). Inspect all parts for wear, scratches or scoring and renew if necessary. Use a feeler gage to check rotor clearances as shown in Figs. 102, 103 and 104. Refer to the following specification data and renew pump assembly if rotor wear limits are exceeded.

Fig. 107 — Exploded view of engine oil pump assembly typical of 1510 and 1710 models.

1. Oil transfer tube
2. Oil pressure relief valve
3. Drive gear
4. End cover
5. Shaft
6. Inner rotor
7. Pin
8. Outer rotor
9. "O" ring
10. Pin
11. Pump housing
12. "O" ring
13. Pickup tube

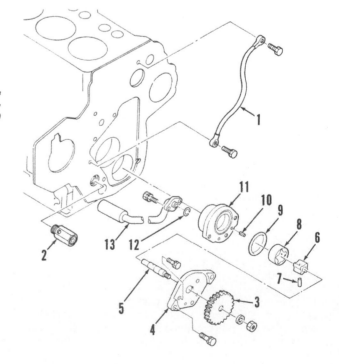

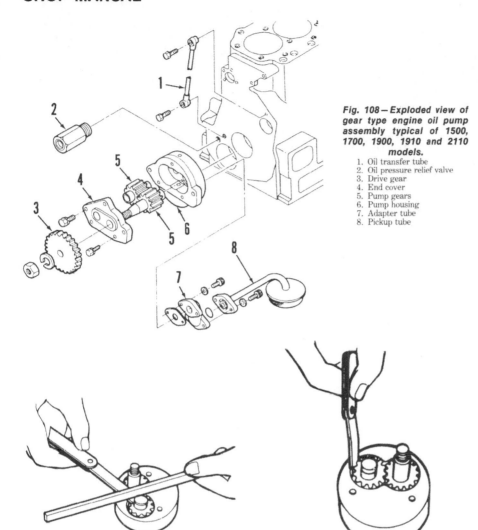

Fig. 108 — Exploded view of gear type engine oil pump assembly typical of 1500, 1700, 1900, 1910 and 2110 models.

1. Oil transfer tube
2. Oil pressure relief valve
3. Drive gear
4. End cover
5. Pump gears
6. Pump housing
7. Adapter tube
8. Pickup tube

brate an injection pump or reset an injector without proper specifications, equipment and training.

FUEL FILTER

All Models

55. Fuel filter life depends upon careful maintenance as well as the hours of operation. The necessity for careful filling with clean fuel cannot be overstressed.

Fuel filter sediment bowl (Fig. 112 or Fig. 113) should be drained whenever water or sediment is visible in bowl. Fuel filter element should be renewed after every 200 hours of operation, or sooner if loss of engine power is evident. After installing new filter, open fuel shut-off valve and loosen vent screws. Tighten vent screws after bowl is filled and all air is expelled from fuel system.

56. **BLEEDING.** The fuel system should be bled if fuel tank is allowed to run dry or if fuel lines, filter or other components within system have been removed. To bleed air from system, first make sure that tank has sufficient amount of fuel. Open fuel shut-off valve and loosen bleed screws (Fig. 112, 113 or 114). Tighten bleed screws when fuel without bubbles flows from around the screws.

If engine fails to start after completing bleeding procedure, loosen high pressure fuel lines at injectors. Move throttle to high speed position, then crank engine with starter until fuel escapes from loosened connections. Tighten fuel line connections and start engine.

Fig. 109—Use a straightedge and feeler gage to measure end clearance between face of pump gears and machined surface housing. Refer to text.

Fig. 110—Measure clearance between gear teeth and housing using a feeler gage as shown. Refer to text.

DIESEL FUEL SYSTEM

INJECTION PUMP

The maintenance of absolute cleanliness is of the utmost importance when servicing the injection pump. Service work or disassembly of injection pump other than that specified should not be attempted without necessary equipment and training.

On 1100, 1110, 1200 and 1300 models, the two pumping elements of fuel injection pump are operated by a camshaft which is located in the timing gear case. On 1210 and 1310 models, the three pumping elements of the injection pump are operated by lobes on the engine valve train camshaft. The fuel injection pump used on all other models is also a multiple plunger pump, but the pump camshaft is located in the injection pump housing.

Because of extremely close tolerances and precise requirements of all diesel components, it is of utmost importance that only clean fuel is used and careful maintenance be practiced at all times. Unless necessary special tools are available, service on injectors and injection

pumps should be limited to removal, installation and exchange of complete assemblies. It is impossible to recali-

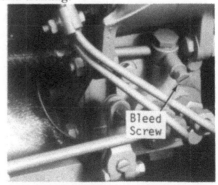

Fig. 112 — View of fuel filter and air bleed screw used on Model 1110. Models 1100, 1200 and 1300 are similar.

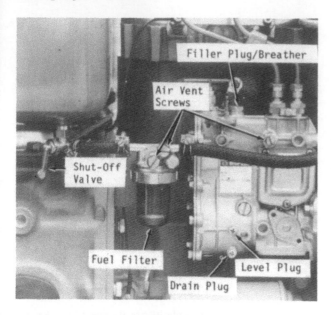

Fig. 113—View of fuel system filter and air bleed locations typical of 1500, 1700, 1710, 1900, 1910 and 2110 models. Oil level plug and oil drain plug are not used on 1700, 1910 and 2110 models.

Fig. 114—View of air bleed screw on Model 1510. Models 1210 and 1310 are similar.

The injection pump on 1500, 1700, 1710, 1900, 1910 and 2110 models is lubricated by the same type of oil as is used in the engine. On 1500, 1700 and 1900 models, oil in pump should be drained and pump filled with new oil to level plug opening after every 200 hours of operation. Oil drain plug, level plug and fill plug are shown in Fig. 113. On 1710, 1910 and 2110 models, the injection pump utilizes oil from the engine crankcase for lubrication. When changing engine oil (every 100 hours of operation), add approximately 150 mL (5 ounces) of clean oil through pump filler opening for 1710 models or approximately 235 mL (8 ounces) of oil for 1910 and 2110 models.

Models 1100-1110-1200-1210-1300-1310

57. **REMOVE AND REINSTALL.** To remove injection pump, first shut off fuel supply. Disconnect fuel inlet line and remove injector lines. Plug all openings to prevent entry of dirt. Set pump throttle lever in midposition. Remove pump mounting cap screws. On 1210 and 1310 models, raise pump and discon-

nect governor control link from pump control rack pin. On all models, remove pump being careful not to damage or lose shims located between pump and timing gear case or cylinder block. The shims are used to adjust injection pump static timing.

To reinstall pump, reverse the removal procedure. If pump is being renewed or if timing shim thickness is not known, refer to paragraph 58 for timing of pump and selection of shims. Be sure pump control rack pin engages governor control link. Tighten pump mounting cap screws to 14-19 N·m (10-14 ft.-lbs.) torque. If delivery valve holders loosened during removal of injector lines, tighten holders to 39-45 N·m (29-33 ft.-lbs.) torque. Bleed air from system as outlined in paragraph 56.

58. **PUMP TIMING.** The injection pump timing must be checked if any of the following components are renewed: injection pump assembly, injection pump camshaft, timing gears or timing gear case. Timing is adjusted by changing thickness of shims located between injection pump and timing gear case or cylinder block mounting surface.

To check pump timing, shut off fuel supply to pump. Disconnect injector line from front fitting (delivery valve holder) on pump. Remove delivery valve holder, delivery valve and spring (Fig. 115). Reinstall delivery valve holder and tighten snugly.

Rotate crankshaft clockwise until No. 1 piston is on compression stroke and TDC mark (2–Fig. 116) on crankshaft pulley is aligned with timing pointer (1). Then, turn crankshaft counterclockwise approximately 30°. Turn on fuel supply to pump and note that fuel should be flowing from No. 1 delivery valve holder. Slowly turn crankshaft clock-

wise to locate exact point at which fuel stops flowing from delivery valve holder, which is start of injection. Pump timing should be correct if the first timing mark (3) on crankshaft pulley is aligned with pointer (1) at this point.

The following pump spill timing procedure may also be used if there is any doubt about accuracy of crankshaft pulley timing marks. Remove valve rocker cover, then rotate crankshaft until No. 1 piston is on compression stroke and timing mark (2) on crankshaft pulley is aligned with timing pointer. Remove valve spring from one of the No. 1 cylinder valves and allow valve to rest on top of the piston. Position a dial indicator on top of valve stem as shown in Fig. 117. Locate No. 1 piston at TDC and zero the dial indicator, then turn crankshaft counterclockwise until dial indicator reading is approximately 7.0 mm (0.275 inch).

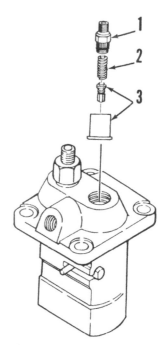

Fig. 115—Exploded view of delivery valve assembly typical of 1100, 1110, 1200 and 1300 models. Other models are similar.

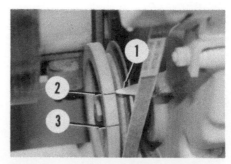

Fig. 116—View of timing pointer (1) and crankshaft pulley showing top dead center timing mark (2) and start of injection timing mark (3).

Fig. 117—Use a dial indicator (1) on top of valve stem (2) to locate No. 1 piston at beginning of injection position when checking injection pump timing. Refer to text.

Turn on fuel supply to pump and move pump control lever to full speed position. Note that fuel should flow from No. 1 delivery valve holder. Slowly turn crankshaft clockwise to locate exact point at which fuel stops flowing from delivery valve holder. The correct timing position (dial indicator reading) at which fuel stops flowing on 1100, 1200 and 1300 models is between 4.055-4.407 mm (0.160-0.174 inch), which is 23°-24° BTDC. On 1110 models, dial indicator reading should be between 3.393-3.717 mm (0.134-0.146 inch), which is 21°-22° BTDC. On 1210 and 1310 models, dial indicator reading should be between 2.850-3.137 mm (0.112-0.124 inch), which is 20°-21° BTDC.

If pump timing is incorrect, increase or decrease shim thickness between pump and timing gear case or cylinder block. Changing shim thickness 0.10 mm (0.004 inch) will change timing approximately 1°. Adding to shim thickness retards timing (and raises piston position), while decreasing thickness of shims will advance timing (and lower piston position).

Do not exceed 0.60 mm (0.012 inch) shim thickness. If more shims are required, refer to paragraph 32 and make sure that timing gears are correctly assembled. If correct timing is possible only with all shims removed, use a thin coat of sealer (such as Loctite 515) to seal pump mounting surface.

Tighten pump mounting screws to a torque of 14-19 N·m (10-14 ft.-lbs.). Remove delivery valve holder and reinstall delivery valve and spring. Tighten delivery valve holder to a torque of 39-45 N·m (29-33 ft.-lbs.)

Model 1510

60. **REMOVE AND REINSTALL.** The injection pump (1–Fig. 120) can be removed separately from governor case (5) if desired. Shut off fuel, disconnect fuel inlet line and remove injector lines. Plug all openings to prevent entry of dirt. Remove intake manifold. Remove pump mounting cap screws, then lift pump assembly from governor case.

To remove governor case with governor asembly and pump camshaft, first remove pump drive gear cover plate from timing gear case. Rotate crankshaft until timing marks (T–Fig. 121) on pump drive gear and idler gear are aligned. Remove drive gear retaining

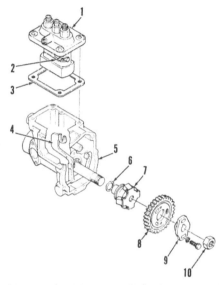

Fig. 120—Exploded view of injection pump and governor case assembly used on Model 1510.

1. Injection pump	6. Shim
2. Control rack pin	7. Flyweight assy.
3. Gasket	8. Drive gear
4. Governor arm	9. Coupling
5. Governor case	10. Retaining nut

Fig. 121—Align timing marks (S and T) on pump drive gear (1), idler gear (2) and coupling (3) as shown when installing injection pump on 1510 models. Keyway (4) should be positioned counterclockwise from coupling timing mark (S).

nut and withdraw drive coupling (9–Fig. 120), drive gear (8), flyweight assembly (7) and shim (6) from injection pump shaft. Disconnect throttle linkage rod from governor lever. Remove cap screws attaching governor case to engine front plate and withdraw governor case.

To reinstall governor case and injection pump, reverse the removal procedure while noting the following special instructions. Make certain that pump control rack pin (2) engages slot in governor arm (4). Be sure that timing marks (T–Fig. 121) on pump drive gear are aligned and that drive coupling to gear timing marks (5) are aligned. If injection pump, governor case, pump camshaft, drive coupling or timing gears are renewed, check pump timing as outlined in paragraph 61. Tighten drive coupling retaining nut to 147-157 N·m (109-115 ft.-lbs.) torque. Tighten coupling to drive gear cap screws to 25-29 N·m (19-22 ft.-lbs.) torque. Bleed air from system as outlined in paragraph 56.

61. **PUMP TIMING.** The injection pump timing must be checked if any of the following components are renewed: injection pump assembly, injection pump camshaft, governor case, drive coupling or timing gears.

To check pump timing, shut off fuel supply and disconnect No. 1 (front) injector line from delivery valve holder on pump. Remove the No. 1 delivery valve holder (1–Fig. 115), spring (2) and delivery valve piston (3). Reinstall delivery valve holder and tighten snugly.

Rotate crankshaft clockwise until No. 1 piston is on compression stroke and TDC timing mark (2–Fig. 116) on crankshaft pulley is aligned with timing pointer (1). Then, turn crankshaft counterclockwise approximately 30° BTDC. Turn on fuel supply to pump and move pump control lever to full speed position. Slowly turn crankshaft clockwise while observing fuel flowing from No. 1 delivery valve holder. Stop turning crankshaft at exact point at which fuel stops flowing from delivery valve holder, which is start of injection. Pump timing should be correct if the first timing mark (3) on crankshaft pulley is aligned with pointer (1) at this point.

The following pump spill timing procedure may also be used if there is any doubt about accuracy of crankshaft pulley timing marks. Remove valve rocker cover, then rotate crankshaft until No. 1 piston is on compression stroke and timing mark (2) on crankshaft pulley is aligned with timing pointer. Remove valve spring from one of the No. 1 cylinder valves and allow valve to rest on top of piston. Position a dial indicator on top of valve stem as shown in Fig. 117. Locate No. 1 piston at TDC and zero the dial indicator, then turn crankshaft counterclockwise until dial indicator reading is approximately 7.0 mm (0.275 inch). Turn on fuel supply to pump and move throttle control lever to high speed position. Note that fuel should flow from No. 1 delivery valve holder. Slowly turn crankshaft clockwise to locate exact point at which fuel stops flowing, which is start of injection. Pump timing is correct if dial indicator reading is between 3.55-3.88 mm (0.140-0.153 inch) at this point. This is specified timing setting of 21½°-22½° BTDC.

If pump timing is incorrect, remove pump drive gear cover plate from timing gear case. Loosen two cap screws attaching pump drive coupling (3–Fig. 121) to drive gear (1). Rotate drive coupling fully counterclockwise in slotted mounting holes while holding drive gear from moving. Make sure that crankshaft is positioned at correct setting as outlined above. Turn on fuel and observe fuel flowing from No. 1 delivery valve holder. While applying counterclockwise pressure on pump drive gear to remove backlash, slowly move drive coupling clockwise to locate exact point at which fuel stops flowing from delivery valve holder. Tighten drive coupling retaining cap screws to 25-29 N·m (19-22 ft.-lbs.) torque being careful not to disturb timing setting.

Note that pump drive coupling to drive gear timing marks (S–Fig. 121), if present, should be aligned when pump is correctly timed. If timing marks are not present, or if they are no longer aligned, use a chisel to mark the coupling and gear for future timing reference.

Reinstall delivery valve and spring. Tighten delivery valve holder to 39-44 N·m (29-32 ft.-lbs.) torque.

Models 1500-1700-1710-1900-1910-2110

62. REMOVE AND REINSTALL. To remove injection pump, proceed as follows: On all models except 1710, drain engine coolant and remove radiator. On 1500, 1700 and 1900 models, remove timing gear case as outlined in paragraph 35. On 1710, remove the injection pump gear cover plate from timing gear case. On 1910 and 2110 models, remove hydraulic pump and filter as an assembly, then remove cover plate from front of timing gear case. On all models, rotate crankshaft until timing marks on pump drive gear and idler gear are aligned as shown in Fig. 123. Make an alignment mark (S) on oil pump drive coupling and gear if no marks are present. Remove retaining nut, drive gear and coupling from pump shaft.

Fig. 122—Scribe alignment marks (5) on pump flange and mounting plate before removing injection pump on 1500, 1700, 1710, 1900, 1910 and 2110 models. (Model 2110 shown.)

1. Fuel inlet line
2. Throttle rod
3. Injector lines
4. Hydraulic pump assy.
5. Alignment marks
6. Injection pump assy.

Disconnect fuel inlet line (1–Fig. 122) and injector lines (3). Plug all openings to prevent entry of dirt. Disconnect throttle control rod (2). Scribe alignment marks (5) on pump mounting flange and mounting plate to assist in installation of pump. Remove pump mounting screws and withdraw injection pump assembly.

Installation of injection pump is the reverse of removal procedure. However, if injection pump, drive coupling or timing gears are renewed, time pump to engine as outlined in paragraph 63. Tighten injection pump gear retaining nut to 39-44 N·m (29-33 ft.-lbs.) torque. Bleed air from system as outlined in paragraph 56.

63. PUMP TIMING. The injection pump timing must be checked if the injection pump or any of the drive components are renewed.

To check pump timing, shut off fuel supply and disconnect No. 1 (front) injector line from delivery valve holder on pump. Remove No. 1 delivery valve holder (1–Fig. 115), spring (2) and delivery valve piston (3), then reinstall delivery valve holder and tighten snugly.

Rotate crankshaft clockwise until No. 1 piston is on compression stroke and TDC mark (2–Fig. 116) on crankshaft pulley is aligned with timing pointer (1). Then, turn crankshaft counterclockwise approximately 30°. Turn on fuel supply to pump and move throttle lever to full speed position. Note that fuel should flow from No. 1 delivery valve holder. Slowly turn crankshaft clockwise to locate exact point at which fuel stops flowing, which is start of injection. Pump timing should be correct if the first mark (3) on crankshaft pulley is aligned with pointer (1) at this point.

The following pump spill timing procedure may also be used to check pump timing if there is any doubt about accuracy of crankshaft pulley timing marks. On 1500, 1700 and 1900 models, the following procedure must be used when installing a new pump and adjusting timing. Remove valve rocker cover. Rotate crankshaft until No. 1 piston is on compression stroke and timing mark (2) on crankshaft pulley is aligned with pointer (1). Remove valve spring from one of the No. 1 cylinder valves and allow the valve to rest on top of the piston. Position a dial indicator on top of the valve stem as shown in Fig. 117. Locate No. 1 piston at TDC and zero the dial indicator, then turn crankshaft counterclockwise until dial indicator reading is approximately 7.0 mm (0.275 inch). Turn on fuel supply and move throttle control lever to high speed position. Note that fuel should flow from No. 1 delivery valve holder. Slowly turn crankshaft clockwise to locate exact point at which fuel stops flowing and observe dial indicator reading. Pump timing is correct if dial indicator reading is within the following specified range.

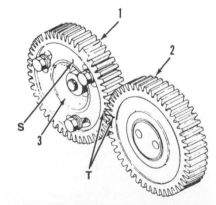

Fig. 123—Align timing marks (S and T) on pump drive gear (1), idler gear (2) and drive coupling (3) when installing injection pump on 1500, 1700, 1710, 1900, 1910 and 2110 models.

Model	Indicator Reading	Crankshaft Degrees
1500...	5.1-6.0 mm (0.200-0.236 in.)	23-24
1700...	3.87-4.67 mm (0.152-0.183 in.)	20-22
1710...	4.12-4.49 mm (0.162-0.177 in.)	22½-23½
1900...	5.24-5.60 mm (0.206-0.220 in.)	26-27
1910...	5.28-5.73 mm (0.208-0.225 in.)	23½-24½
2110...	5.28-5.73 mm (0.208-0.225 in.)	23½-24½

If pump timing is incorrect, adjust as follows: On 1500, 1700 and 1900 models, remove radiator, hydraulic pump and timing gear case. On 1710 models, remove pump drive gear cover plate from timing gear case. On 1910 and 2110 models, remove radiator, hydraulic pump and pump drive gear cover plate. On all models, loosen cap screws attaching pump drive coupling (3–Fig. 123) to pump drive gear (1), then rotate coupling counterclockwise to end of slotted holes in drive gear. Make certain that crankshaft is positioned at correct setting as outlined above. Turn on fuel and observe fuel flowing from No. 1 delivery valve holder. While applying counterclockwise pressure on pump drive gear to remove backlash, slowly turn pump drive coupling clockwise to locate exact point at which fuel stops flowing. At this point, tighten drive coupling retaining cap screws to 25-29 N·m (19-22 ft.-lbs.) torque being careful not to disturb timing setting.

Note that pump drive coupling to drive gear timing marks (S–Fig. 123) should be aligned when pump is correctly timed. If timing marks are not present, or if they are no longer aligned, use a chisel to mark the coupling and gear for future timing reference.

Reinstall delivery valve and spring. Tighten delivery valve holder to 30-35 N·m (22-26 ft.-lbs.) torque.

ENGINE SPEED ADJUSTMENT

Models 1100-1110-1200-1300

65. To adjust engine speed, first start engine and operate until warm. Move throttle lever to slow idle detent position. Engine speed should be 750-850 rpm. If low idle speed is incorrect, loosen locknuts and turn throttle control rod turnbuckle (3–Fig. 125) as required until correct speed is obtained.

Move throttle control lever to maximum speed position and observe engine rpm. Maximum no-load speed should be 2750-2800 rpm for 1100, 1110 and 1200 models, or 2900-2950 rpm for 1300 models. To adjust maximum speed,

loosen locknut and turn stop bolt (Fig. 125) as required until correct maximum speed is obtained.

Models 1210-1310-1500-1510-1700-1710-1900-1910-2110

66. To adjust engine speed, first start engine and operate until warm. Move throttle lever to slow idle detent position. Engine speed should be 750-850 rpm on all models. If low idle speed is incorrect, loosen locknuts and turn throttle control rod turnbuckle (Fig. 126) as required until correct speed is obtained.

Move throttle control lever to maximum speed position. Maximum no-load speed should be as follows:

Model	High Idle Rpm
1210	2850-2900
1310	2950-3000
1500	2650-2700
1510	3000-3050
1700	2600-2650
1710	2825-2875
1900	2900-2950
1910	2650-2700
2110	2650-2700

To adjust maximum speed, turn control arm stop screw (1–Fig. 127, 128 or 129) as required until correct speed is obtained. Be sure that foot throttle pedal does not travel below the upper surface of foot step plate.

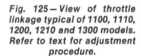

Fig. 125 – View of throttle linkage typical of 1100, 1110, 1200, 1210 and 1300 models. Refer to text for adjustment procedure.

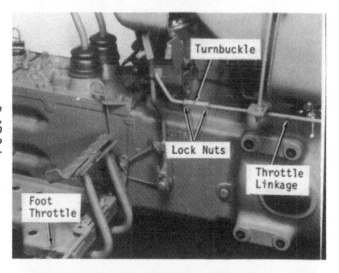

Fig. 126 – View of throttle linkage typical of 1310, 1500, 1510, 1700, 1710, 1900, 1910 and 2110 models. Refer to text for adjustment procedure.

Fig. 127 – View of throttle linkage adjustment points typical of 1210 and 1310 models.
1. High idle stop screw
2. Shut-off stop screw

Fig. 128 — View of throttle linkage adjustment points for 1510 models.

1. High idle stop screw
2. Shut-off stop screw

Fig. 129 — View of throttle linkage adjustment points typical of 1500, 1700, 1710, 1900, 1910 and 2110 models.

1. High idle stop screw
2. Shut-off stop screw

Engine should shut off quickly when throttle control lever is moved to STOP position. If engine continues to run, turn control arm stop screw (2) counterclockwise as necessary until engine shuts off when control lever is moved to STOP position.

GOVERNOR

Models 1100-1110-1200-1300

70. The governor flyweights are located on the forward end of engine camshaft. The governor control linkage is located inside timing gear case. To service governor linkage or flyweights, refer to paragraph 33 for removal of timing gear case. Refer to Fig. 132 for exploded view of flyweights and associated parts. Refer to Fig. 133 for exploded view of linkage.

When reassembling linkage, tighten screws (7—Fig. 134) while positioning components of control lever arm to provide the following dimensions. **Models 1100, 1200 and 1300:** Dimension (A—

Fig. 134A), measured between lower fork arm (8) and gasket surface of timing gear case, should be 26 mm (1.024 in.) with throttle in closed (low idle) position. Dimension (B—Fig. 134A), measured between center of slot in governor arm (10) and gasket surface of timing gear case, should be 48 mm (1.890 in.). **Model 1110:** Dimension (A—Fig. 134A) should be 27.4 mm (1.078 in.) and dimension (B) should be 57.5 mm (2.265 in.) with throttle in closed position.

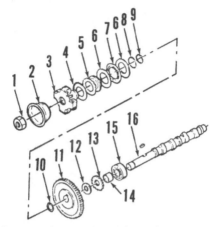

Fig. 132 — Governor flyweights and related components are located on end of engine camshaft on 1100, 1110, 1200 and 1300 models.

1. Nut
2. Oil slinger
3. Flyweight assy.
4. Shim
5. Slider
6. Washers
7. Thrust bearing
8. Snap ring
9. Snap ring
10. Shim
11. Camshaft gear
12. Spacer
13. Tachometer gear
14. Collar
15. Bearing
16. Camshaft

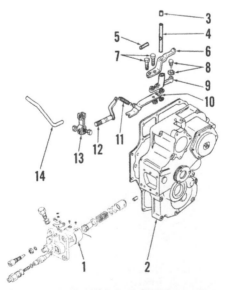

Fig. 133 — Exploded view of timing gear case showing governor linkage typical of 1100, 1110, 1200 and 1300 models.

1. Injection pump
2. Timing gear case
3. Plug
4. Shaft
5. Pin
6. Arm
7. Bolts
8. Pin and follower
9. Arm
10. Arm
11. Spring
12. Arm and shaft
13. External lever
14. Speed control rod

Models 1210-1310

71. The governor flyweights are located on the forward end of engine camshaft. The governor control linkage is located inside the timing gear case. To service governor linkage or flyweights, refer to paragraph 34 for removal of timing gear case. Refer to Fig. 135 for an exploded view of flyweight and associated parts and to Fig. 136 for an exploded view of governor linkage.

When reassembling linkage, align groove in governor arm shaft with slot of governor lever as shown at (A—Fig. 137).

Model 1510

72. The governor flyweights and related components are located on the end of injection pump camshaft. The governor linkage is located in governor housing. To service governor linkage or flyweights, refer to paragraph 60 for removal of injection pump and governor housing. Refer to Fig. 138 for an exploded view of governor components.

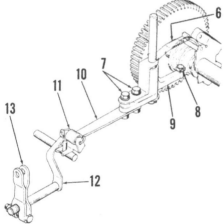

Fig. 134 — Drawing of governor weights and linkage typical of 1100, 1110, 1200 and 1300 models. Refer to Fig. 133 for legend.

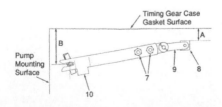

Fig. 134A — Refer to text for correct positioning of control lever arms (6, 9 and 10) before tightening bolts (7).

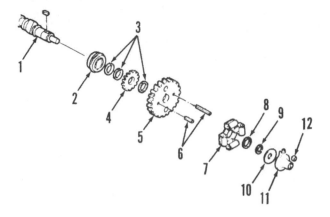

1. Camshaft
2. Bearing
3. Spacers
4. Tachometer drive gear
5. Camshaft gear
6. Drive pins
7. Flyweight assy.
8. Shim
9. Snap ring
10. Shim
11. Slider
12. Collar

paragraphs for removal and test procedures.

76. **REMOVE AND REINSTALL.** Before removing injector lines and injectors, thoroughly clean injectors and surrounding area. Remove fuel leak-off lines and high pressure lines. Unscrew nozzle assembly from cylinder head.

Precombustion chambers (Fig. 140) are fitted in cylinder head on 1100, 1110, 1200, 1210, 1300, 1310, 1510 and 1710 models. The glow plugs must be removed before attempting to remove the prechambers. A special tool (Nuday No. 1589) is available for removing prechamber retaining ring (4). Remove cylinder head and tap out prechambers from underside of head, or use a suitable puller and slide hammer to remove prechambers.

Install injectors using new gaskets. Tighten injectors to a torque of 59-69 N·m (43-51 ft.-lbs.).

Models 1500-1700-1710-1900-1910-2110

73. The governor assembly is located on rear end of the injection pump camshaft within the injection pump housing. The injection pump and governor should be serviced only by an authorized diesel injection service center.

INJECTOR NOZZLES

All Models

75. **TESTING AND LOCATING A FAULTY NOZZLE.** If rough or uneven engine operation, or misfiring indicates a faulty nozzle, the defective unit can usually be located as follows:

With engine running at the speed where malfunction is most noticeable, loosen compression nut on high pressure line for each injector in turn allowing fuel to escape at the nut rather than enter the cylinder. If engine operation is not affected when injector line is loosened, that is the cylinder that is misfiring.

If a faulty injector is found and considerable time has elapsed since injectors have been serviced, it is recommended that all injectors be removed and serviced, or that new or reconditioned units be installed. Refer to the following

77. **TESTING.** A complete job of testing and adjusting injector nozzles requires use of special test equipment. Nozzle should be tested for opening pressure, seat leakage and spray pattern. When tested, nozzle should open with a buzzing sound, and cut off quickly at end of injection.

WARNING: Fuel leaves injector nozzle with sufficient force to penetrate the skin. Keep exposed portions of your body clear of nozzle spray when testing.

Before conducting tests, operate tester lever until fuel flows, then attach injector to tester. Operate tester lever a few quick strokes to purge air from injector and to make sure that nozzle valve is not stuck.

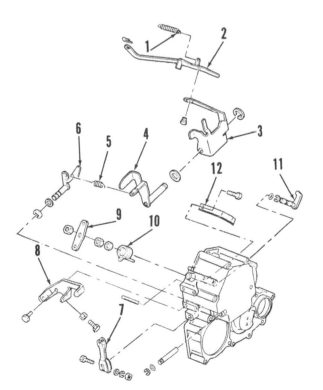

Fig. 136—Exploded view of governor linkage used on 1210 and 1310 models.

1. Spring
2. Lever
3. Lever
4. Arm
5. Spring
6. Arm
7. Control lever
8. Stop bolt bracket
9. Spring
10. Spring
11. Arm
12. Leaf spring

Fig. 137—On 1210 and 1310 models, align groove in governor arm shaft with slot in control lever as shown at (A).

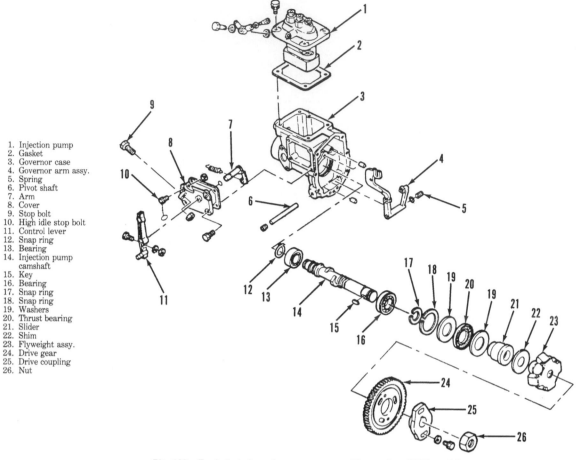

1. Injection pump
2. Gasket
3. Governor case
4. Governor arm assy.
5. Spring
6. Pivot shaft
7. Arm
8. Cover
9. Stop bolt
10. High idle stop bolt
11. Control lever
12. Snap ring
13. Bearing
14. Injection pump camshaft
15. Key
16. Bearing
17. Snap ring
18. Snap ring
19. Washers
20. Thrust bearing
21. Slider
22. Shim
23. Flyweight assy.
24. Drive gear
25. Drive coupling
26. Nut

Fig. 138 — Exploded view of governor assembly used on 1510 models.

78. OPENING PRESSURE. Operate tester lever slowly while observing tester gage reading. Opening pressure should be 11720 kPa (1700 psi) for all models. Opening pressure is adjusted by adding or removing shims in shim pack (8 – Fig. 141).

79. SPRAY PATTERN. The spray pattern should be well atomized and slightly conical, emerging in a straight axis from nozzle tip. If pattern is wet, ragged or intermittent, nozzle must be overhauled or renewed.

80. SEAT LEAKAGE. Wipe nozzle tip dry, then operate tester lever to bring gage pressure to 1035 kPa (150 psi) below opening pressure and hold this pressure for 10 seconds. If any fuel appears on nozzle tip, overhaul or renew nozzle.

81. OVERHAUL. Hard or sharp tools, emery cloth, grinding compound or other than approved solvents or lapping compounds must never be used. An approved nozzle cleaning kit is available through a number of specialized sources.

Wipe all dirt and loose carbon from exterior of nozzle and holder assembly.

Refer to Fig. 141 for exploded view and proceed as follows:

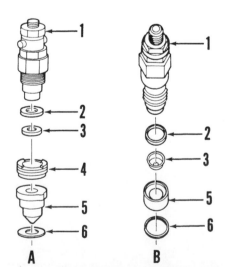

Fig. 140 — Precombustion chamber assembly shown at (A) is fitted in cylinder head of 1100, 1110, 1200, 1300, 1510 and 1710 models. Precombustion chamber shown at (B) is used on 1210 and 1310 models.

1. Injector assy.
2. Seal washer
3. Heat shield
4. Retaining ring
5. Prechamber
6. Seal washer

Secure nozzle in a soft jawed vise or holding fixture and remove nut (3). Place all parts in clean calibrating oil or diesel fuel as they are removed. Use a compartmented pan to keep parts from each injector together and separate from other units if more than one injector is being serviced.

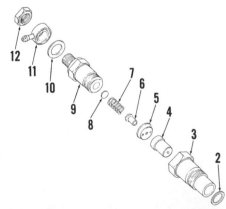

Fig. 141 — Exploded view of typical injection nozzle.

2. Gasket
3. Nozzle nut
4. Nozzle
5. Plate
6. Pin
7. Spring
8. Shim
9. Nozzle holder
10. Gasket
11. Banjo fitting
12. Nut

Clean exterior surfaces with a brass wire brush, soaking in an approved carbon solvent if necessary, to loosen hard carbon deposits. Rinse parts in clean diesel fuel or calibrating oil immediately after cleaning to neutralize the solvent and prevent etching of polished surfaces.

Clean nozzle spray orifice hole using 1.0 mm (0.040 inch) wire in a pin vise. Scrape carbon from pressure chamber using hooked scraper. Clean valve seat using brass scraper.

Reclean all parts by rinsing thoroughly in clean diesel fuel or calibrating oil and assemble while parts are immersed in cleaning fluid. Make sure adjusting shim pack is intact. Tighten nozzle retaining nut (3) to a torque of 61-75 N·m (45-55 ft.-lbs.). Do not overtighten, distortion may cause valve to stick and no amount of overtightening can stop a leak caused by scratches or dirt. Retest assembled injector as previously outlined.

GLOW PLUGS
All Models

82. Glow plugs are parallel connected with each glow plug grounding through mounting threads. Start switch is provided with a "HEAT" position which can be used to energize the glow plugs for faster warm up. If "COLD START AID" indicator light fails to glow when start switch is held in "HEAT" position an appropriate length of time (approximately 30 seconds), check for loose connections at switch, indicator lamp, glow plug connections and ground. A test lamp can be used at glow plug connection to check for current to glow plug.

COOLING SYSTEM

All models use a pressurized cooling system which raises coolant boiling point. An impeller type centrifugal pump is used to provide forced circulation. A thermostat is used to stabilize operating temperature.

RADIATOR

All Models

85. Radiator cap pressure valve is set to open at 90 kPa (13 psi) on all models. It is recommended that a 50/50 mix of ethylene glycol base antifreeze and water be used for coolant. Refer to CONDENSED SERVICE DATA section for cooling system capacities.

To remove radiator, first drain the coolant. Remove upper and lower radiator hoses. Remove air cleaner hose if necessary. Unbolt and remove radiator assembly.

THERMOSTAT

All Models

86. The thermostat is located in the coolant outlet elbow on all models except 1210 and 1310 models. On 1210 and 1310 models, the thermostat is located in the water pump housing. On all models, thermostat should begin to open at 71° C (160° F) and be completely open at 85° C (185° F).

WATER PUMP

Models 1100-1110-1200-1300

87. **R&R AND OVERHAUL.** To remove water pump, first drain coolant from radiator and cylinder block. Loosen alternator attaching bolts and remove fan belt. Remove radiator and cooling fan. Remove water pump mounting bolts and remove the pump.

To disassemble, pull drive pulley (10 – Fig. 142) off pump shaft. Support pump housing, then press shaft and bearing assembly (8) forward out of impeller (5) and pump housing. Remove the seal assembly (6).

When assembling water pump, apply nonhardening sealer to outer circumference of seal (6). Install seal into housing by pressing carefully on outer edge of seal. Install shaft and bearing assembly (8) into housing by pressing on outer edge of bearing until front end of bearing is flush with front of housing as shown at (A – Fig. 143). Press impeller (5) onto shaft until surfaces (B) are flush as shown. Support impeller and shaft surfaces (B), then press drive hub (10) onto shaft until distance (D) is 36 mm (1.417 inches) 1110 models, or 39.5 mm (1.555 inches) for 1100, 1200 and 1300 models.

Installation of pump is reverse of removal procedure.

Models 1210-1310

88. **R&R AND OVERHAUL.** To remove water pump, first drain coolant from radiator and cylinder block. Loosen alternator mounting bolts and remove fan belt. Remove radiator as outlined in paragraph 85. Remove fan and fan pulley. Remove pump mounting bolts and withdraw water pump assembly.

Remove mounting plate (4 – Fig. 144), spring (1) and thermostat (2). Press shaft and bearing assembly (9) forward out of impeller (6) and housing (8). Press shaft out of front hub (10) if necessary. Press seal (7) rearward from pump housing.

When assembling, apply nonhardening sealer to outer circumference of new seal. Install seal in housing by pressing against outer edge of seal until it bottoms against housing shoulder. Install shaft and bearing asembly into housing by pressing against outer edge of bearing until front of bearing is flush with

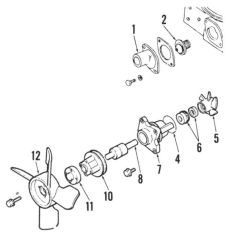

Fig. 142 – Exploded view of water pump used on 1100, 1110, 1200 and 1300 models. Thermostat (2) is also shown.

1. Thermostat housing	7. Housing
2. Thermostat	8. Shaft & bearing assy.
4. "O" ring	10. Pulley hub
5. Impeller	11. Spacer
6. Seal assy.	12. Fan

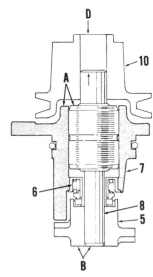

Fig. 143 – Cross section of water pump used on 1100, 1110, 1200 and 1300 models. Refer to text for assembly dimensions (A, B and D) and to Fig. 142 for legend.

front of housing as shown at (A–Fig. 145). Press impeller onto shaft until end of shaft is flush with rear of impeller (B). Support impeller and shaft surfaces (B), then press hub onto shaft until flush with front of shaft as shown at (C).

Installation of pump is reverse of removal.

Models 1500-1510-1700-1710-1900-1910-2110

89. **R&R AND OVERHAUL.** To remove water pump, first drain coolant from radiator and cylinder block. Remove radiator as outlined in paragraph 85. Loosen alternator mounting bolts, then remove fan belt and fan. Remove pump mounting bolts and remove water pump assembly.

Remove mounting plate (3–Fig. 146). Remove set screw (9) from housing. Press shaft and bearing assembly (8) forward out of impeller (5) and housing. Press seal (6) rearward out of housing. Press pump shaft out of pulley (10).

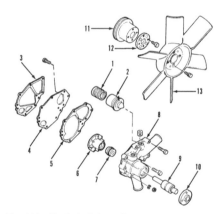

Fig. 144—Exploded view of water pump assembly used on 1210 and 1310 models. Thermostat (2) is located in pump housing (8).

1. Spring
2. Thermostat
3. Gasket
4. Mounting plate
5. Gasket
6. Impeller
7. Seal assy.
8. Housing
9. Shaft & bearing assy.
10. Hub
11. Pulley
12. Spacer
13. Fan

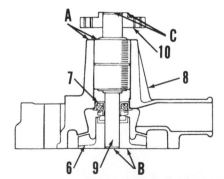

Fig. 145 – Cross section of water pump used on 1210 and 1310 models. Refer to text for assembly dimensions (A, B and C) and to Fig. 144 for legend.

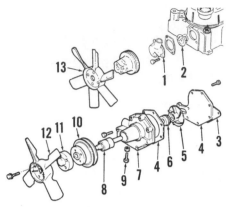

Fig. 146—Exploded view of water pump typical of type used on 1500, 1700 and 1900 models. Water pump used on 1510, 1710, 1910 and 2110 models is similar.

1. Thermostat housing
2. Thermostat
3. Plate
4. Gaskets
5. Impeller
6. Seal assy.
7. Housing
8. Shaft & bearing assy.
9. Set screw
10. Pulley
11. Spacer
12. Fan (1500)
13. Fan

When assembling water pump, apply nonhardening sealer to outer circumference of new seal. Install seal by pressing against outer edge of seal until it bottoms against housing shoulder. Install shaft and bearing assembly into housing by pressing against outer edge of bearing until front of bearing is flush with front surface of pump housing as shown at (A–Fig. 147). Be sure that groove around center of bearing is aligned with set screw hole, then install

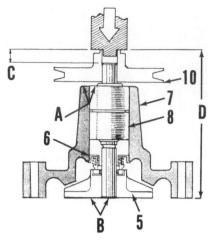

Fig. 147 – Cross section of water pump typical of 1500, 1510, 1700, 1710, 1900, 1910 and 2110 models. Refer to text for assembly dimensions (A, B, C and D) and to Fig. 146 for legend.

set screw (9) and jam nut. Press impeller (5) onto shaft until rear face of impeller is flush with end of shaft as shown at (B). Support impeller and shaft surface (B), then press pulley (10) onto shaft until the following dimensions are obtained: On 1510 and 1710 models, dimension (C) should be 11.2-11.4 mm (0.441-0.449 inch). On 1500, 1700 and 1900 models, dimension (D) should be 167.8-168.2 mm (6.606-6.622 inches). On 1910 and 2110 models, dimension (D) should be 198.3 mm (7.807 inches).

Installation of water pump is reverse of removal.

ELECTRICAL SYSTEM

ALTERNATOR AND REGULATOR

All Models

90. A Nippon Denso alternator and regulator are used on 1100, 1110, 1200, 1210 and 1300 models. A Hitachi alternator and regulator are used on all other models. The voltage regulator is located on a plate just above steering gear base on 1100, 1110, 1200 and 1210 models. On 1300, 1500 and 1700 models, the regulator is mounted on fuel tank separator plate in engine compartment. On all other models, the regulator is located on a support bracket under the instrument panel.

When servicing the electrical system, the following precautions must be observed to avoid damage to charging circuit components.

a. When installing battery or connecting a booster battery, the negative post must be grounded.

b. Always disconnect battery ground cable before removing or installing any electrical components.

c. Do not connect or disconnect any charging circuit wiring when engine is running.

d. Never short across any terminal of alternator or regulator unless specifically recommended.

91. **OUTPUT TEST.** To check alternator output, disconnect output wire at alternator and connect a test ammeter in series with alternator output terminal and wire as shown in Fig. 150. Connect a load tester and voltmeter to battery terminals.

Turn on headlights for one minute, then record battery voltage. Start the engine and set engine speed at approximately 1600 rpm. Using the load tester, apply load to charging system to obtain maximum alternator output and record voltmeter and ammeter readings. Alternator may be rated at either 20 amps or 35 amps.

If voltmeter reading increases but remains below 15.5 volts (normal regulated voltage is 13.8-14.8) and ammeter reading is within 10 percent of rated amps, charging system is operating properly. If voltmeter reading exceeds 15.5 volts, regulator is faulty. If voltmeter reading remains the same or decreases and ammeter reading is less than specified, perform the following maximum field output test.

Connect a jumper wire (3–Fig. 151) between alternator output terminal (4) and field terminal (1) prior to starting engine. Then start engine and perform output test using same procedure as outlined above.

If voltmeter or ammeter reading remains the same or decreases, alternator is faulty. If voltmeter or ammeter reading increases, regulator or wiring is faulty. (There must be current at alternator "F" terminal when key switch is turned to "ON" position before alternator will begin charging.)

If alternator is charging but indicator light remains "ON," check alternator "N" circuit voltage as follows: With engine running, connect voltmeter positive lead to "N" terminal (2–Fig. 151) on alternator and connect voltmeter negative terminal to ground. Voltmeter reading should be 4.0-5.8 volts for 1100, 1110, 1200, 1210 and 1300 models and 4.2-5.2 volts for all other models. If "N" voltage is less than specified minimum, alternator is faulty. If "N" voltage is within specified range, regulator is faulty.

92. **OVERHAUL.** Refer to appropriate Fig. 152, 153 or 154 for an exploded view of alternator. Prior to disassembly, scribe matching marks across both end housings and stator frame for reference when assembling. Remove through-bolts and separate front housing and rotor from rear housing and stator. Remove attaching nuts and separate stator from rear housing. When unsoldering stator wires from diode rectifier, use a preheated soldering iron and separate connections as

quickly as possible to avoid heat damage to diodes. Rotor can be removed from front housing after removing retaining nut, drive pulley and fan.

Inspect all parts for wear, burned or discolored wiring or other damage and renew if necessary. Using an ohmmeter, check for open, shorted or grounded circuits as follows:

Touch one tester lead to each of the rotor slip rings. Resistance should be 4.2 ohms for 1100, 1110, 1200, 1210 and 1310 models; 10.6 ohms for 1500, 1700 and 1900 models; 3.9 ohms for 1310, 1510 and 1710 models; 3.8 ohms for 1910 and 2110 models. Excessively low or high resistance reading indicates shorted or open circuit, and rotor should be renewed. Renew rotor if there is continuity (grounded circuit) between slip rings and rotor frame.

There should be continuity between each of the stator wire leads. There should not be continuity between winding leads and stator frame. Renew stator if windings are discolored or burned.

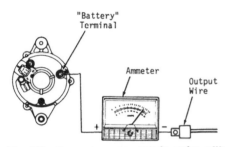

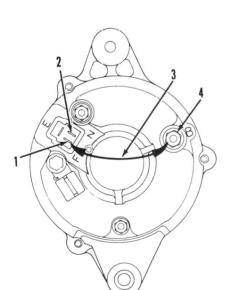

Fig. 150—Connect an ammeter in series with alternator "battery" terminal and output wire as shown to check alternator current output.

Fig. 151—To check alternator maximum field output, connect a jumper wire (3) between alternator output terminal (4) and field terminal (1) before starting engine. Refer to text.

Fig. 152—Exploded view of Nippon Denso alternator typical of type used on some models.

1. Pulley
2. Fan
3. Front cover
4. Spacer
5. Washer, cover & bearing
6. Retainer plate
7. Rotor
8. Bearing
9. Stator
10. Diode plate
11. Brushes & holder
12. Housing

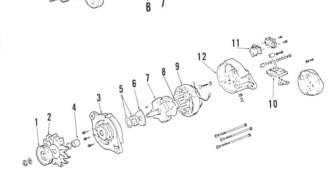

Fig. 153—Exploded view of Hitachi alternator typical of type used on some models.

1. Pulley
2. Fan
3. Front cover
4. Spacer
5. Washer, cover & bearing
6. Retainer plate
7. Rotor
8. Bearing
9. Stator
10. Diode plate
11. Brushes & holder
12. Housing

Fig. 154—Exploded view of Hitachi alternator typical of type used on some models. Refer to Fig. 153 for legend.

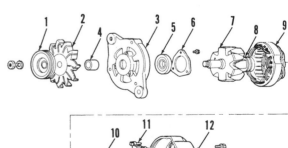

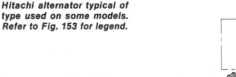

Check rectifier diodes by touching one tester lead to diode terminal and touch other lead to rectifier frame, then reverse test lead connections. The diodes should show continuity in one direction and high resistance in the other direction. If one or more diodes has high resistance in both directions or low resistance in both directions, renew rectifier assembly.

Refer to the following specifications for rotor and brush wear limits.

Models 1100-1110-1200-1210-1300

Rotor Slip Rings
  Minimum Diameter . . . . . . .31.7 mm
                         (1.248 in.)
  Maximum Runout . . . . .0.05 mm
                         (0.002 in.)
Brush Length –
  Minimum. . . . . .5.5 mm
                         (0.217 in.)

Models 1310-1510-1710

Rotor Slip Rings –
  Minimum Diameter . . . . .30 mm
                         (1.181 in.)
  Maximum Runout . . . . .0.05 mm
                         (0.002 in.)
Brush Length –
  Minimum. . . . . .8.5 mm
                         (0.335 in.)

Models 1500-1700-1900

Rotor Slip Rings –
  Minimum Diameter . . . . . . .30.7 mm
                         (1.209 in.)
  Maximum Runout . . . . .0.05 mm
                         (0.002 in.)
Brush Length –
  Minimum . . . . .7 mm
                         (0.275 in.)

Models 1910-2110

Rotor Slip Rings –
  Minimum Diameter . . . . . . .31.3 mm
                         (1.232 in.)
  Maximum Runout . . . . .0.05 mm
                         (0.002 in.)
Brush Length –
  Minimum . . . . .9 mm
                         (0.355 in.)

When reassembling alternator, solder stator wire connections using resin core solder. Hold rectifier terminals with needlenose pliers to absorb heat during soldering operation. Quickly cool soldered connection with a damp cloth to protect rectifier diodes. To hold brushes in retracted position during assembly, push brushes into holder and insert a pin or wire through holes in rear housing and brush holder.

SAFETY START SWITCH

Models 1100-1110-1200-1210-1300-1500-1700-1900

93. A safety start switch (1 – Fig. 155) is located on left side of clutch housing and is actuated by the clutch pedal. The switch should prevent starter motor from engaging unless clutch pedal is fully depressed. Clutch linkage should be adjusted correctly before adjusting safety start switch. Adjust position of switch so that it passes current to energize starter solenoid only when clutch pedal is fully depressed.

Models 1310-1510-1710-1910-2110

94. The safety start system consists of a magnetic reed switch located on transmission shift cover and a relay switch located under the instrument panel. The magnetic reed switch is actuated by magnets on the main gear shift rails and prevents starter solenoid from engaging unless main shift lever is in neutral.

Shift cover must be removed for access to magnetic reed switch. For proper operation of reed switch, clearance (C – Fig. 156) between reed switch

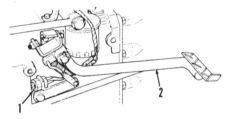

Fig. 155 – Safety start switch (1) is located on left side of clutch housing on 1100, 1110, 1200, 1210, 1300, 1500, 1700 and 1900 models. Switch is actuated by the clutch pedal (2).

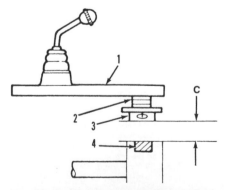

Fig. 156 – Safety start switch (3) is located on shift cover on 1310, 1510, 1710, 1910 and 2110 models. Switch is actuated by magnets in main shift rails.

1. Shift cover   3. Reed switch
2. Shim washers   4. Shift rail magnet

(3) and shift rail magnet (4) must be 3.5-4.0 mm (0.138-0.157 inch). The clearance is adjusted by placing shim washers (2) between transmission cover and the reed switch plate.

STARTING MOTOR AND SOLENOID

Models 1100-1110-1200-1210-1310-1510

95. Refer to appropriate Fig. 157 or 158 for an exploded view of starter and solenoid. Clearance (C – Fig. 159) between pinion and thrust collar should be 0.2-1.5 mm (0.008-0.060 inch) for all models. To check clearance, first disconnect and insulate field coil wire from solenoid terminal. Engage solenoid by attaching jumper cables as shown at (1, 2 and 3). Do not leave solenoid engaged too long or overheating may result. Push pinion back with thumb to remove slack and use a feeler gage to measure clearance (C). Clearance can be adjusted by adding or removing shims (S – Fig. 157) between solenoid and drive housing.

Refer to the following specifications:

Model 1210

Armature Shaft Runout –
  Maximum . . . . .0.08 mm
                         (0.003 in.)
Commutator Diameter –
  Minimum . . . . .28 mm
                         (1.102 in.)
Commutator Insulation Depth –
  Minimum 0.2mm
                         (0.008 in.)
Brush Length –
  Minimum . . . . .11.5 mm
                         (0.452 in.)
No-Load Bench Test –
  Current Draw . . . . .50 amps
  Rpm 5000
Load Test . . . . .270 amps

Models 1100-1110-1200-1310-1510

Armature Shaft Runout –
  Maximum . . . . .0.08 mm
                         (0.003 in.)
Commutator Diameter –
  Minimum . . . . .43 mm
                         (1.693 in.)
Commutator Insulation Depth –
  Minimum. . . . . .0.2 mm
                         (0.008 in.)
Brush Length –
  Minimum . . . . .12 mm
                         (0.472 in.)
No-Load Bench Test –
  Current Draw . . . . .60 amps
  Rpm 6000
Load Test . . . . .540 amps

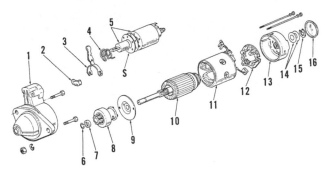

Fig. 157—Exploded view of starting motor and solenoid typical of 1100, 1110, 1200, 1310 and 1510 models.

1. Housing
2. Seal
3. Lever
4. Spring
5. Solenoid
6. Retaining ring
7. Stop
8. Pinion
9. Center plate
10. Armature
11. Field housing
12. Brushes & plate
13. Cover
14. Washers
15. Retaining ring
16. Cover

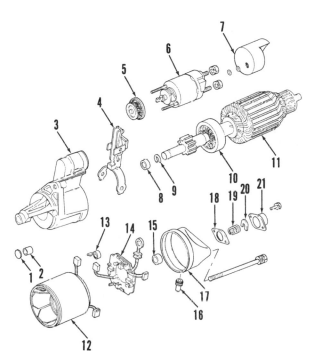

Fig. 158—Exploded view of starting motor and solenoid used on 1210 models.

1. Plug
2. Bushing
3. Drive housing
4. Lever
5. Cover
6. Solenoid
7. Cover
8. Thrust collar
9. Retaining ring
10. Clutch & pinion assy.
11. Armature
12. Field housing
13. Spring
14. Brush holder assy.
15. Bushing
16. Plug
17. End frame
18. Gasket
19. Spring
20. Retainer clip
21. Cover

Model 1300

96. Refer to Fig. 160 for an exploded view of starting motor. The starter solenoid is enclosed in housing (5). To check solenoid, connect wires (1, 2 and 3—Fig. 161) to locations indicated. The pinion should extend from housing. Disconnect wire (3) from terminal and pinion should remain extended. If pinion does not extend with all three jumper wires connected or if pinion does not remain extended when wire (3) is disconnected, renew solenoid assembly.

Refer to the following specifications:

Armature Shaft Runout –
  Maximum . . . . 0.1 mm
    (0.004 in.)
Commutator Diameter –
  Minimum . . . . 29 mm
    (1.142 in.)
Commutator Insulation Depth –
  Minimum . . . . 0.4 mm
    (0.016 in.)
Brush Length –
  Minimum . . . . 12 mm
    (0.472 in.)
No-Load Bench Test
  Current Draw . . . . 90 amps
  Rpm 3500
Load Test . . . . 400 amps

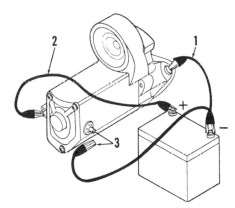

Fig. 161—Refer to text for checking starter solenoid on 1300 models.

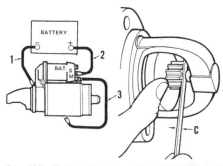

Fig. 159—Pinion clearance (C) can be checked by connecting jumpers as shown between battery ground and starter housing (1), solenoid "S" terminal and battery positive terminal (2) and "M" terminal of solenoid to starter motor ground (3). Refer to text.

Fig. 160—Exploded view of starting motor used on 1300 models.

1. Housing
2. Seal
3. Plunger rod
4. Spring
5. Solenoid assembly
6. Clutch
7. Bearing
8. Bearing
9. Washer
10. Armature
11. Field coil housing
12. Brushes
13. Retainer
14. Rollers
15. Gear
16. Gear
17. Ball
18. Bearing
19. Bearing

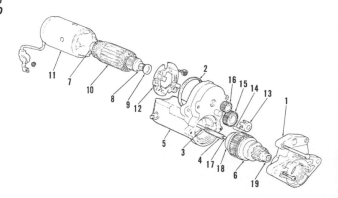

Models 1500-1700-1900-1910-2110

97. Refer to Fig. 162 for an exploded view of starting motor and solenoid. Pinion clearance (C—Fig. 159) should be 0.2-1.5 mm (0.008-0.060 inch) with solenoid engaged. To check clearance, disconnect and insulate field wire from solenoid terminal. Engage solenoid by attaching jumper cables (1, 2 and 3) to locations indicated. Push pinion back to remove slack and use a feeler gage to measure clearance between pinion and stop collar. Clearance can be adjusted by adding or removing shims (S—Fig. 162) between solenoid and drive housing.

Refer to the following specifications:

Armature Shaft Runout —
Maximum . . . . .0.1 mm
(0.004 in.)
Commutator Diameter —
Minimum . . . . .40 mm
(1.575 in.)
Commutator Insulation Depth —
Minimum. . . . . .0.2 mm
(0.008 in.)
Brush Length – Minimum
1500 and 1700 Models . . . . . .11.5 mm
(0.452 in.)
All Other Models . . . . .15 mm
(0.590 in.)

No-Load Bench Test —
Current Draw . . . . .90 amps
Rpm 4000
Load Test
1500 and 1700 Models . . . . .900 amps
All Other Models . . . . .1300 amps

Model 1710

98. Refer to Fig. 163 for an exploded view of starting motor and solenoid. Pinion shaft end play (P—Fig. 164) should

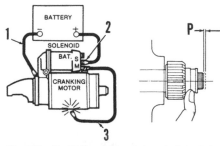

Fig. 164—On 1710 models, pinion shaft end play (P) can be checked after connecting jumper wires as shown between battery ground and starter housing (1), solenoid "S" terminal and battery positive terminal (2), and solenoid "M" terminal to starter housing.

be 0.5-2.0 mm (0.020-0.080 inch). To check shaft end play, disconnect and insulate field coil wire from solenoid terminal. Engage solenoid by attaching jumper wires (1, 2 and 3) to locations indicated. Push pinion gear back to remove overtravel, then measure shaft movement. Add or remove shims (7—Fig. 163) to obtain recommended end play.

Refer to the following specifications:

Armature Shaft Runout —
Maximum . . . . .0.05 mm
(0.002 in.)
Commutator Diameter —
Minimum . . . . .32 mm
(1.260 in.)
Commutator Insulation Depth –
Minimum. . . . . .0.2 mm
(0.008 in.)
Brush Length —
Minimum . . . . .11 mm
(0.433 in.)
No-Load Bench Test —
Current Draw . . . . .130 amps
Rpm 4000

ENGINE CLUTCH

ADJUSTMENT

All Models

100. Clutch pedal free travel should be 20 to 30 mm (¾ to 1-1/16 inches). Free travel is measured at pedal pad as indicated in Fig. 165 or 166. To adjust, disconnect clevis (2) from bellcrank and lengthen or shorten rod (1) as required to obtain recommended free travel.

On models equipped with dual clutch, adjustment of pto clutch release bolts can be checked after removing rubber cover from side of clutch housing. Clearance between head of bolts (3—Fig. 167) and pto pressure plate (4) should be between 1.6-1.8 mm (0.063-0.070 inch) for 1500, 1700, and 1900 models or 0.9-1.0 mm (0.035-0.039 inch) for 1310, 1510, 1710, 1910 and 2110 models.

Check adjustment of safety start switch (3—Fig. 165), if so equipped, as outlined in paragraph 93 after clutch linkage is adjusted.

CLUTCH SPLIT

All Models

101. To separate (split) tractor between engine and clutch housing, drain oil from transmission and rear axle

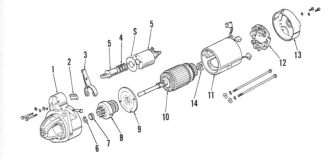

Fig. 162 — Exploded view of starting motor typical of type used on 1500, 1700, 1900, 1910 and 2110 models.

1. Housing
2. Seal
3. Lever
4. Spring
5. Solenoid
6. Retaining ring
7. Stop
8. Pinion
9. Center plate
10. Armature
11. Field coil housing
12. Brushes & plate
13. Cover
14. Washers

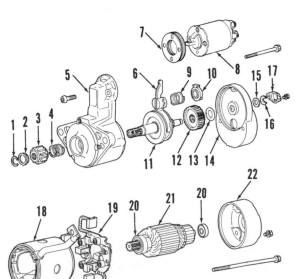

Fig. 163 — Exploded view of starting motor and solenoid used on 1710 models.

1. Retaining ring
2. Stop collar
3. Pinion gear
4. Spring
5. Drive end housing
6. Fork
7. Shims
8. Solenoid
9. Spring
10. Guide
11. Pinion shaft
12. Gear
13. Shim
14. Center housing
15. Shim
16. Retainer clip
17. Cover
18. Field frame
19. Brush holder assy.
20. Bearings
21. Armature
22. End Cover

Fig. 165 — View of clutch linkage typical of 1100, 1110, 1200 and 1210 models. Pedal free travel should be 20-30 mm (3/4 to 1-3/16 inches).

a suitable hoist or rolling jack. Remove cap screws attaching engine to clutch housing, then move engine away from clutch housing.

CLUTCH ASSEMBLY

Single Plate Clutch

102. **R&R AND OVERHAUL.** The clutch and pressure plate can be removed from flywheel after separating the engine from clutch housing as described in paragraph 101. Loosen clutch mounting screws evenly to reduce chances of distorting pressure plate.

Refer to appropriate Fig. 170, 171 or 172 for exploded view of clutch com-

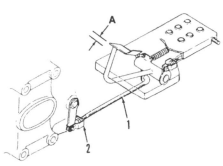

Fig. 166 — View of clutch linkage typical of 1300, 1310, 1500, 1510, 1700, 1710, 1900, 1910 and 2110 models. Pedal free travel (A) should be 20-30 mm (3/4 to 1-3/16 inches).

Fig. 170 — Exploded view of clutch and release bearing used on 1100 and 1200 models. Models 1110 and 1210 are similar.
1. Friction disc
2. Pressure plate assy.
3. Release bearing
4. Bearing hub
5. "O" ring
6. Fork
7. Pins
8. Support housing
9. Oil seals
11. Band
12. Rubber boot
13. Snap ring
14. Bushing
15. Bushing
16. Cross shaft
17. Clutch housing

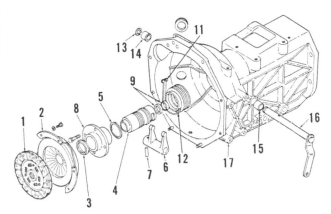

Fig. 167 — On models with dual plate clutch, pto clutch release bolts (3) can be adjusted through opening in clutch housing. Refer to text.

Fig. 171 — Exploded view of single plate clutch and release bearing typical of 1300, 1310, 1510 and 1710 models.
1. Friction disc
2. Pressure plate assy.
3. Release bearing
4. Bearing hub
6. Fork
7. Pin
8. Support housing
9. Oil seal
13. Snap ring
14. Bushing
15. Bushing
16. Cross shaft
17. Clutch housing
18. Return spring

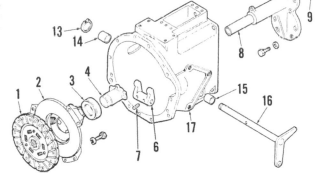

housings. Remove front end weights, if equipped, and install wedges between frame and front axle to prevent engine from tipping. Remove the hood. Disconnect battery ground cable, then disconnect wiring to headlights, glow plugs, temperature sender, oil pressure sender, alternator and starter. Remove the starter motor assembly. Disconnect tachometer cable. Detach steering drag link from steering arm. Shut off fuel and disconnect fuel supply line and injector leak-off line. Disconnect throttle linkage from governor control lever. Disconnect hydraulic lines and power steering line if so equipped. Place a support stand under clutch housing. Support engine with

Fig. 172 — Exploded view of single plate clutch and release bearing typical of 1500, 1700 and 1900 models. Refer to Fig. 171 for legend.

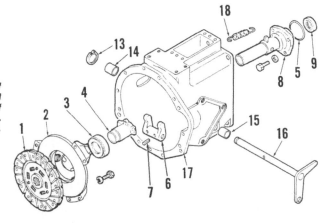

ponents. Inspect clutch for wear or damage and renew if necessary. Minor imperfections can be removed from pressure plate by resurfacing friction surface.

NOTE: Friction surface of a new pressure plate is coated with a protective film which must be removed using a suitable solvent prior to installation.

When installing clutch, longer side of clutch disc hub should face away from flywheel. Use a suitable pilot tool to align clutch disc while attaching clutch to flywheel. Tighten retaining screws evenly.

Release lever height above surface of flywheel should be 54 mm (2.125 inches) on 1310, 1500, 1510, 1700, 1710 and 1900 models. Release lever height is not adjustable on other models.

Dual Plate Clutch

103. R&R AND OVERHAUL. The dual plate clutch assembly can be removed from flywheel after separating engine from clutch housing as described in paragraph 101. Loosen clutch mounting cap screws evenly to prevent distortion. Refer to Figs. 173 and 174 for exploded view of clutch components.

The use of special clutch fixture tool (No. 2706) is recommended for releasing spring tension on clutch release levers during disassembly and reassembly of clutch and for setting release lever height (Fig. 175).

When assembling clutch, be sure pto clutch disc (D) is installed with longer side of hub facing away from flywheel. Clearance (C) between head of pto clutch release bolt and pto pressure plate should be between 1.6-1.8 mm (0.063-0.070 inch) for 1500, 1700 and 1900 models or 0.9-1.0 mm (0.035-0.039

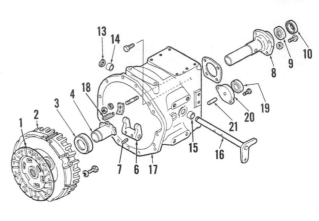

Fig. 173—Exploded view of dual plate clutch and release bearing typical of 1310, 1500, 1510, 1700, 1710, 1900, 1910 and 2110 models. Refer to Fig. 174 for exploded view of clutch assembly components.

1. Friction disc
2. Clutch assy.
3. Release bearing
4. Bearing hub
6. Fork
7. Pins
8. Support housing
9. Oil seal
10. Bearing
13. Snap ring
14. Bushing
15. Bushing
16. Cross shaft
17. Clutch housing
18. Return spring
19. Bearing
20. Retainer
21. Pin

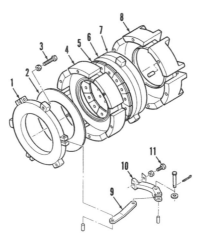

Fig. 174—Exploded view of dual plate clutch used on some models.

1. Transmission pressure plate
2. Diaphragm spring
3. Adjusting bolt
4. Inner cover
5. Pto clutch disc
6. Pto pressure plate
7. Diaphragm spring
8. Outer cover
9. Link
10. Release lever
11. Adjusting bolt

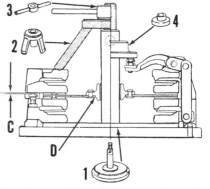

Fig. 175—Special clutch fixture tool is used to disassemble, reassemble and set clutch release lever height. Refer to text.

inch) for 1310, 1510, 1710, 1910 and 2110 models.

When installing a new pressure plate assembly, clean pressure plate friction surface using suitable solvent to remove protective film. Use a suitable pilot tool to align clutch discs while attaching clutch to flywheel. Tighten clutch mounting bolts gradually and evenly to 25-34 N·m (18-25 ft.-lbs.) torque.

STANDARD TRANSMISSION

(1100-1110-1200-1210 Models)

The standard transmission has five unsynchronized forward gears and one reverse gear combined with a high-low range sliding gear to provide ten forward speeds and two reverse speeds. An optional creeper gear is available which provides fifteen forward speeds and three reverse speeds.

INSPECTION

All Models So Equipped

105. Removal of the shift cover (65 – Fig. 176 or 177) will permit visual inspection of most of the transmission components. Remove dipstick (67), then

remove screws attaching cover (65) to transmission housing. Carefully lift cover away from housing.

Gears and shafts can be better inspected through top opening after draining fluid and flushing system with appropriate solvent. Transmission, center housing (differential and final drive), rear axles and hydraulic system share a common sump. Check filters and drained fluid for evidence of metal particles. Refer to paragraph 106 for lubrication requirements.

LUBRICATION

All Models So Equipped

106. Transmission, center housing (differential and final drive), rear axles and hydraulic system share a common sump. The drain plugs are located at bottom of axle housings (Fig. 178). The fluid should be drained and refilled with Ford 134 or equivalent fluid every 300 hours of operation. Capacity is approximately 18.9 liters (20 U.S. quarts) for 1100 and 1200 models or 17 liters (18 U.S. quarts) for 1110 and 1210 models. Fluid level should be maintained at full mark on dipstick located in shift cover. Fill opening is also located in shift cover.

The hydraulic system oil filter is located by the oil pump on left side of engine. The filter should be cleaned after every 300 hours of operation. Renew filter element if it is damaged or cannot be cleaned satisfactorily.

REMOVE AND REINSTALL

All Models So Equipped

107. To remove transmission, first drain transmission, center housing, rear axles and hydraulic system fluid. Disconnect foot throttle linkage and brake rods. Remove brake pedals, brake cross shaft and foot platforms. Disconnect battery ground cable, then disconnect wiring to rear of tractor. Disconnect

hydraulic lines from hydraulic manifold on right side of tractor, and remove hydraulic pump suction line from left side of transmission housing. Remove transmission shift cover (65 – Fig. 176 or 177), high-low shifter lever and cover (48) and detent plug, spring and ball (50).

Support the center (differential) housing at front and rear to prevent tipping. Install wedges between front axle and frame rails to prevent engine from tipping. Place rolling floor jack under clutch housing. Support transmission with an overhead hoist. Remove cap screws attaching clutch housing to transmission, then move engine and clutch housing away from transmission. Remove cap screws attaching transmission housing

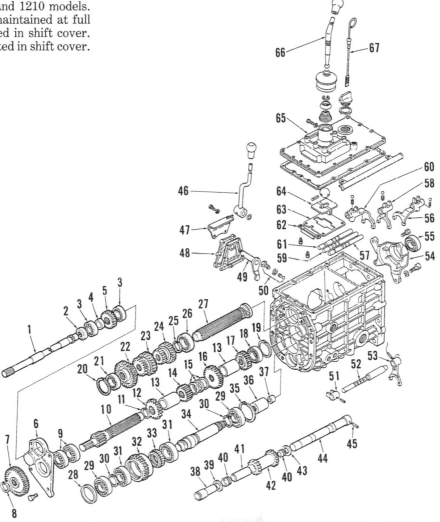

Fig. 176 – Exploded view of standard transmission housing and gears typical of 1100 and 1200 models.

1. Input shaft	14. Gear (2nd)	24. Sliding gear	35. Snap ring	57. Shift rail
2. Oil seal	15. Shims	(3rd/5th)	36. Coupling	(1st/Reverse)
3. Bearings	16. Gear (5th)	25. Snap ring	37. Alignment bushing	58. Fork (2nd/4th)
4. Spacer	17. Gear (3rd)	26. Bearing	38. Collar	59. Shift rail (2nd/4th)
5. Input gear	18. Bearing	27. Top (counter) shaft	39. Thrust washer	60. Fork (3rd/5th)
6. Front cover	19. Snap ring	28. Spacer	40. Bearings	61. Shift rail (3rd/5th)
7. Input driven gear	20. Snap ring	29. Bearings	41. Spacer	62. Shift guide
8. Snap ring	21. Bearing	30. Snap rings	42. Reverse idler gear	63. Shift plate
9. Bearings	22. Sliding gear	31. Bearings	43. Thrust washer	64. Guide
10. Mainshaft	(1st/Reverse)	32. Pto counter gear	44. Reverse gear shaft	65. Cover
11. Snap ring	23. Sliding gear	33. One-way clutch	45. Pin	66. Shift lever
12. Gear (4th)	(2nd/4th)	34. Pto countershaft	46. Range select lever	67. Dipstick
13. Spacers (2 used)			47. Guide	
			48. Cover	
			49. Shift arm and shaft	
			50. Detent plug, spring	
			and ball	
			51. Shift boss	
			52. Shift shaft	
			53. Shift fork	
			54. Bearing holder	
			55. Bearing	
			56. Fork (1st/Reverse)	

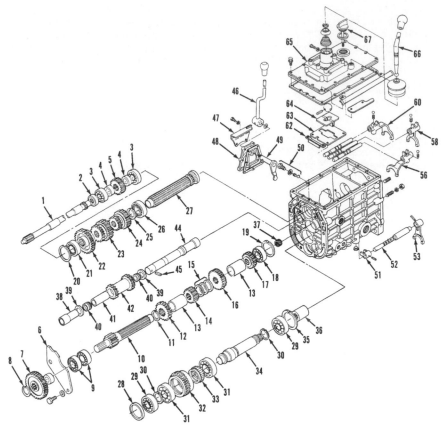

Fig. 177—Exploded view of standard transmission housing and gears typical of 1110 and 1210 models. Refer to Fig. 176 for legend.

to center housing and remove transmission assembly.

NOTE: Be careful not to damage high-low range gears located between rear of transmission housing and front of center housing as transmission is being removed. Reposition gears on differential shaft as housings are separated.

Reinstall transmission by reversing the removal procedure. Apply liquid gasket maker to mating surfaces of transmission and center housings.

OVERHAUL

All Models So Equipped

108. Refer to appropriate Fig. 176 or 177 for an exploded view of transmission assembly. To disassemble, slide shift rails rearward from housing while removing shift forks (56, 58 and 60) with detent springs and balls from the rails. Remove input shaft (1) with bearings (3) and gear (5).

Remove snap ring (8), driven gear (7) and spacer (if used) from front of mainshaft (10). Remove front cover plate (6). Pull countershaft (27) rearward while withdrawing sliding gears (22, 23 and 24) out top of housing. Pull mainshaft

(10) with bearings (9) out front of housing while removing gears, spacers (13) and shims (15).

Remove roll pin (45) from reverse idler shaft (44). Remove reverse idler gear (42), washers (39) and spacer (41) as shaft is withdrawn rearward from housing. Bump the collar (38) forward out of housing bore using a soft hammer.

Remove snap ring (35) retaining pto countershaft rear bearing (29), then tap shaft rearward until rear bearing is free of the housing. Pull rear bearing from the countershaft, then withdraw parts (29 through 34) as an assembly out top of housing.

Remove roll pin and shifter boss (51) from range shift rail (52), then withdraw shift rail and fork (53) from housing. Remove high-low range countershaft, gear and bearings from front of center housing. Remove high-low sliding gear from bevel pinion shaft.

Inspect all parts for wear or damage and renew if necessary. One-way clutch (33) should permit pto counter gear (32) to rotate clockwise (viewed from rear), but should lock gear to shaft when rotating gear counterclockwise. Pto counter gear should be renewed if inside diameter exceeds 57.173 mm (2.251 inches). Pto countershaft should be renewed if area of shaft that contacts one-

way clutch is less than 4.46 mm (1.593 inches) in diameter. Backlash between gears should be 0.20 mm (0.008 inch) with a wear limit of 0.60 mm (0.024 inch).

To reassemble, reverse the disassembly procedure while noting the following special instructions: Parts (29 through 33) should be assembled on pto countershaft (34) before positioning through top opening. Be sure that reverse idler collar (38) is firmly seated in housing bore. Check mainshaft end play using a dial indicator. Install appropriate shims (15) between second gear (14) and fifth gear (16) to obtain recommended end play of 0.08-0.18 mm (0.003-0.007 inch).

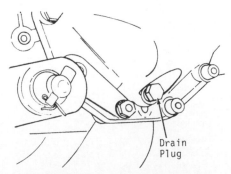

Fig. 178 – Drain plug is located at bottom of axle housing on 1100, 1110, 1200 and 1210 models.

HYDROSTATIC TRANSMISSION

(1110 and 1210 Models)

The hydrostatic transmission consists of a variable displacement piston type hydraulic pump and a fixed displacement piston type hydraulic motor located in the clutch housing. The hydrostatic pump is connected directly to engine clutch disc by the input shaft (1 – Fig. 180). The hydrostatic unit has two output shafts. The upper shaft (4) provides direct drive from engine clutch to pto front countershaft which is located in transmission housing. The lower shaft (6) is hydrostatic motor output shaft which drives the high-low range transmission gears located in transmission housing.

Fig. 180 – View of hydrostatic drive unit used on 1110 and 1210 models.

P. High pressure test ports
1. Input shaft
2. Charge pump
3. Feed valve & neutral valve (2 used)
4. Pto drive shaft
5. High pressure relief valve
6. Transmission drive shaft

LUBRICATION

All Models So Equipped

110. The hydrostatic unit utilizes fluid contained in a common reservoir consisting of transmission housing, center (differential) housing and rear axle housings. The fluid should be drained and refilled with Ford 134 or equivalent fluid every 300 hours of operation. The drain plugs are located at bottom of axle housings (Fig. 178). Capacity is approximately 15.5 liters (16.4 U.S. quarts).

The hydrostatic system is equipped with a suction filter (Fig. 181) and a cartridge type pressure filter. It is recommended that suction filter element be cleaned and cartridge filter renewed after the first 50 hours of operation when tractor is new or whenever hydrostatic unit has been overhauled or renewed. Thereafter, suction filter should be cleaned (or renewed if it is damaged or cannot be cleaned satisfactorily) and cartridge filter renewed every 300 hours of operation.

After suction filter element is reinstalled, remove bleed plug from top of filter cover to allow air to escape from filter body.

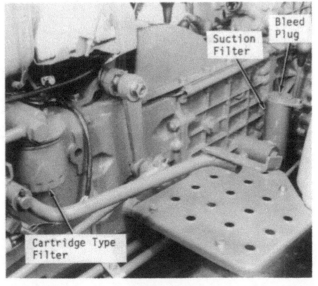

Fig. 181 – Hydrostatic system is equipped with a suction filter and a cartridge type pressure filter.

TROUBLE-SHOOTING

All Models So Equipped

111. The following are some symptoms which may occur during operation of the hydrostatic transmission and their possible causes.

1. Tractor will not move or operates erratically.
   a. Engine clutch or transmission input shaft failure.
   b. Transmission oil level low.
   c. Neutral valve or feed valve faulty.
2. Tractor will move but lacks pulling ability.
   a. High pressure relief valve opening pressure set too low.
   b. Internal leakage of high pressure oil.
3. Abnormal noise when operating.
   a. Suction filter plugged.
   b. Air leak on suction side of charge pump.

c. Charge pump or pump relief valve faulty.

4. Tractor fails to stop at neutral.
a. Control linkage out of adjustment.

5. Oil overheating
a. Hydraulic oil cooler plugged.

TESTS AND ADJUSTMENTS

All Models So Equipped

112. **HIGH PRESSURE RELIEF VALVE.** To check high pressure relief valve setting, first operate tractor until oil temperature is 40°-50° C (100° -120° F). Remove test port pipe plug (P – Fig. 182), which is accessible through opening (D – Fig. 183) in top of clutch housing, and install a 0-35000 kPa (0-5000 psi) pressure gage in test port. Place range lever in high range and securely lock parking brake. Start engine and set speed at 2700 rpm. Slowly move foot control pedal to obtain maximum pressure. Pressure reading should be 21510-25445 kPa (3120-3690 psi).

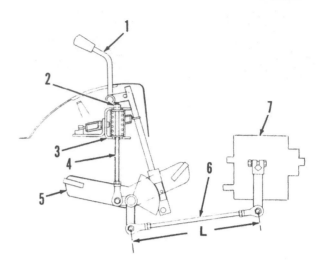

Fig. 184 – Drawing of hydrostatic control linkage showing adjustment points. Length (L) of control rod should be 278 mm (10-15/16 inches).

1. Speed control lever
2. Spring tension adjusting nuts
3. Support bracket
4. Neutral control rod
5. Foot control pedal
6. Control rod
7. Hydrostatic unit

NOTE: Do not maintain maximum pressure any longer than necessary to prevent overheating and possible damage to transmission.

If necessary to adjust the relief valve, the tractor must be separated between engine and clutch housing as outlined in paragraph 101 for access to relief valve (5 – Fig. 180) adjusting screw. If relief valve is faulty, complete port block assembly (37 – Fig. 187) must be renewed.

113. **CHARGE PUMP RELIEF VALVE.** To check charge pressure relief valve setting, first operate tractor until oil temperature is 40°-50° C (100°-120° F). Remove pipe plug from test port (C – Fig. 183) located on left side of filter manifold and install test gage. Start engine and set speed at 2700 rpm. Pressure gage reading should be 490 kPa (71 psi).

Charge relief valve setting is not adjustable. If pressure is low, check for plugged suction filter, faulty relief valve (15 – Fig. 187), faulty charge pump or faulty cylinder block assemblies.

114. **CONTROL LINKAGE ADJUSTMENT.** Adjust length of hydrostatic control rod (6 – Fig. 184) to 278 mm (10-15/16 inches). Adjust neutral rod spring tension nut (2) to provide sufficient tension to return foot pedal (5) to neutral position when released. Loosen support bracket (3) retaining bolts. Shift range lever to low speed position, start engine and operate at low idle speed. Adjust foot control pedal until neutral position is obtained, then tighten support bracket retaining bolts.

Fig. 182 – High pressure test ports (P) are located in top of hydrostatic unit.

Fig. 185 – View of hydrostatic transmission foot control pedal (1) and pedal stop (2).

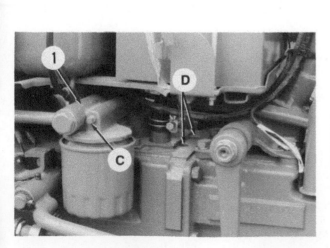

Fig. 183 – View of filter manifold (1) showing location of charge pump pressure test port (C). High pressure test ports are accessible through openings (D) in top of clutch housing.

Fig. 186 – View of hydrostatic drive unit with clutch housing removed.

With engine shut off, depress foot pedal fully in forward position until internal stop is contacted. Adjust pedal stop (2 – Fig. 185) until pedal (1) is moved slightly rearward, then tighten locknut.

REMOVE AND REINSTALL

All Models So Equipped

115. To remove hydrostatic unit, split tractor between engine and clutch housing as outlined in paragraph 101. Remove fuel tank, battery and support bracket, steering gear and instrument panel. Remove suction tube from bottom of hydrostatic unit and remove return tube from top of unit. Disconnect oil cooler inlet and outlet tubes from the filter manifold, then remove the manifold. Unbolt and remove clutch housing from the linkage housing. Remove control lever (1 – Fig. 186) from hydrostatic unit (2). Remove cap screws attaching hydrostatic unit to the linkage housing and remove hydrostatic unit.

To reinstall, reverse the removal procedure.

OVERHAUL

All Models So Equipped

116. To disassemble, remove allen head screws retaining port block (37 – Fig. 187) to hydrostatic case (16). Separate port block from case being careful not to damage valve plates (25 and 31). Unseat snap ring (23) from groove at base of pump cylinder block, then remove pump cylinder block assembly (20 through 24) from hydrostatic case. Remove the motor cylinder block assembly (26 through 31) and output shaft (18) from the case. Scribe alignment marks on swashplate covers (11), then tap end of swashplate trunnion shaft with a soft hammer to remove covers from case. Unbolt and remove support hub (1) and charge pump assembly. Remove input shaft (10) from front of case.

Remove charge pump relief valve (15) from hydrostatic case. Remove neutral valves (32) and feed valves (35) from port block. When removing high pres-

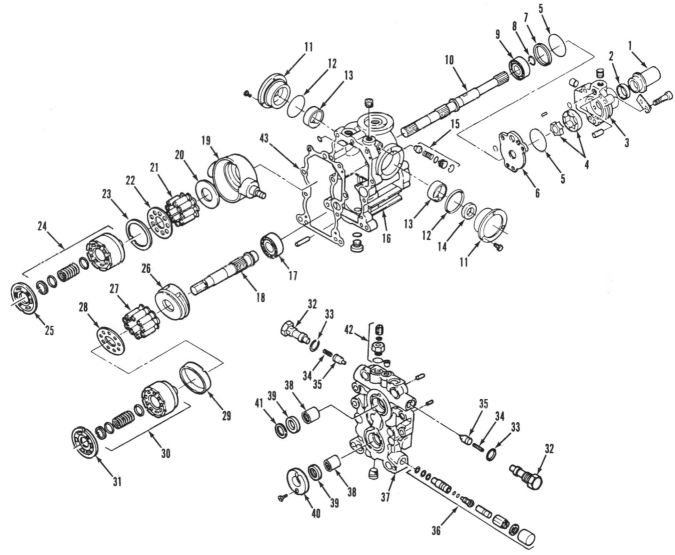

Fig. 187 – Exploded view of hydrostatic drive unit.

1. Clutch release bearing hub	9. Bearing	16. Hydrostatic case	24. Pump cylinder block assy.	31. Valve port plate
2. Oil seal	10. Input shaft	17. Bearing	25. Valve port plate	32. Neutral valve
3. Charge pump housing	11. Cover	18. Output shaft	26. Swash plate	33. "O" ring
4. Pump rotors	12. "O" ring	19. Swash plate	27. Motor pistons	34. Spring
5. "O" rings	13. Bushing	20. Thrust plate	28. Retainer plate	35. Feed valve poppet
6. Wear plate	14. Oil seal	21. Pump pistons	29. Spacer	36. High pressure relief valve
7. Spacer	15. Charge pump relief valve	22. Retainer plate	30. Motor cylinder block assy.	37. Port block
8. Snap ring		23. Snap ring		38. Bearings
				39. Oil seals
				40. Retainer
				41. Retainer
				42. Pressure test port plugs
				43. Gasket

sure relief valve (36), count the number of turns required to remove the pressure adjusting screw so it can be reinstalled to original setting.

Lubricate all parts with clean hydraulic oil during assembly. Port block mounting gasket (43) should be soaked in oil prior to installation. Renew all oil seals.

Install the pump swashplate (19) and covers (11) making sure that cover scribe marks are aligned. Tighten cover retaining screws to 2.0-2.7 N·m (18-24 in.-lbs.) torque. Install input shaft (10), spacer (7), wear plate (6) and charge pump assembly. Tighten charge pump mounting bolts to 16-19 N·m (12-14 ft.-lbs.) torque.

Install pump cylinder block and pistons making sure that snap ring (23) is fully seated in its groove. Install output shaft and motor cylinder block assembly making sure that swashplate (26) is correctly located on dowel pin in bottom of case. Note that face of motor cylinder block will be approximately 5 mm (3/16 inch) below gasket surface of case when properly installed.

Install valve port plates (25 and 31) onto cylinder blocks with brass side facing the cylinder blocks. Note that pump port plate has feathering notches (N – Fig. 188) cut in ends of port passages and is not interchangeable with motor port plate. Install port block (37 – Fig. 187) and tighten mounting bolts evenly to 31-38 N·m (23-28 ft.-lbs.) torque.

RANGE TRANSMISSION

All Models So Equipped

117. **REMOVE AND REINSTALL.** To remove range transmission assembly, first separate engine from clutch housing as outlined in paragraph 101 and remove clutch housing and hydrostatic unit as outlined in paragraph 115. Disconnect foot throttle rod and brake control rods. Remove brake pedals and pedal cross shaft. Remove both foot step plates. Disconnect battery ground cable, then disconnect wiring to rear of tractor. Disconnect hydraulic suction lines and return to sump lines from transmission housing. Remove hydrostatic linkage cover. Disconnect spring (3 – Fig. 190) from speed control brake rod (5) and remove bolts (2) attaching linkage bracket (6) to support bracket. Remove transmission top cover (1). Remove hydraulic lift cover as outlined in paragraph 180. Remove high-low shift lever (1 – Fig. 191) and cover (2) as an assembly. Drive roll pin from shift boss (12) and remove boss from shift rail (11). Remove detent plug (4), spring (5) and ball (6).

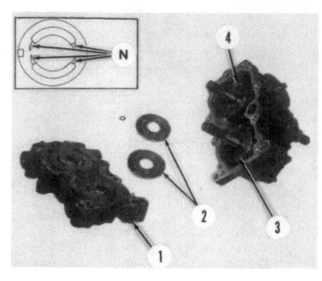

Fig. 188 — Pump valve plate has feathering notches (N) cut on ends of port passages and is not interchangeable with motor valve plate.

1. Port block
2. Valve plates
3. Motor assy.
4. Pump assy.

Fig. 190 — View of hydrostatic transmission control linkage.

1. Range transmission top cover
2. Retaining bolts
3. Spring
4. Speed control lever
5. Brake control rod
6. Linkage bracket
7. Pointer

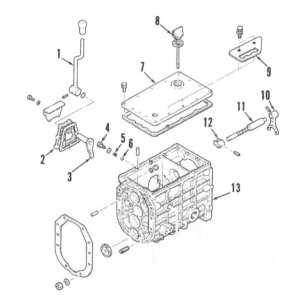

Fig. 191 — Exploded view of hydrostatic drive range transmission shift linkage.

1. High-low shift lever
2. Cover
3. Shift arm
4. Plug
5. Spring
6. Detent ball
7. Top cover
8. Dipstick
10. Shift fork
11. Rail
12. Shift boss
13. Range transmission housing

NOTE: It is recommended that shift rail (11), fork (10) and high-low sliding gears be kept with rear axle center housing when removing transmission housing.

Support rear axle center housing at front and rear to prevent tipping. Install wedges between front axle and side rails to prevent engine from tipping. Support engine and clutch housing with a suitable rolling jack. Attach overhead hoist to range transmisson housing. Remove bolts attaching linkage housing to transmission housing, then move engine and linkage housing away from transmission housing. Remove bolts attaching transmission housing to rear axle center housing. Note that two retaining bolts are located at the top inside transmission housing. Pry transmission housing away from center housing being careful not to damage high-low range gears located between the two housings.

To reinstall transmission, assemble high-low sliding gears, shift fork and rail in rear axle center housing. Position low range gear (9 – Fig. 192), spacer (10) and high range gear (11) on range gear countershaft (6). Slide transmission housing into position on center housing while aligning shift rail with transmission housing bore and bevel pinion shaft with transmission drive countershaft (20). Position shift boss (12 – Fig. 191) on shift rail and install roll pin. Complete installation by reversing the removal procedure.

118. OVERHAUL. To disassemble, remove high-low sliding gears (25 and 27 – Fig. 192), shift fork and shift rail from rear axle center housing. Remove high range gear (11), spacer collar (10), low range gear (9) and snap ring (8) from range gear countershaft (6). Withdraw pto input shaft (4) from front of housing. Remove snap ring (13) and drive gear (14) from front of transmission drive countershaft.

To remove transmission drive countershaft (20A) on early production models (1110 before S.N. UB00785 and 1210 before S.N. UC01417), remove bevel pinion shaft front bearing support (1 – Fig. 193) from rear of transmission housing. Use a brass drift to tap countershaft (20A – Fig. 192) forward until front bearings are free of housing. Remove the bearings (15A) and snap rings (24A) from front of countershaft. Remove snap ring (22A) from rear bearing counterbore, then tap countershaft rearward from housing while withdrawing cluster gear (17A) and thrust washers out top of housing.

To remove transmission drive countershaft (20) on late production models, remove snap ring (22) from rear bearing counterbore. Use a brass drift to tap

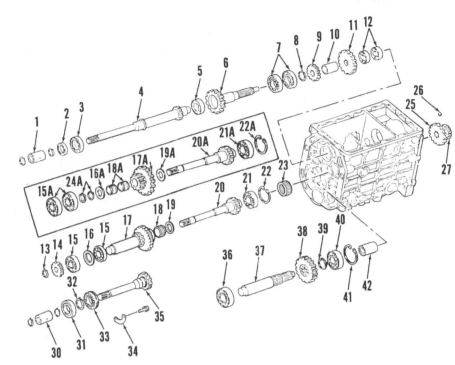

Fig. 192 — Exploded view of range transmission assembly used on 1110 and 1210 models equipped with hydrostatic drive. Countershaft (20A) and related components shown in inset are used on early production models (1110 before S.N. UB00785 and 1210 before S.N. UC 01417).

1. Coupling	12. Bearings	19 & 19A. Thrust washer	30. Coupling
2. Oil seal	13. Snap ring		31. Oil seal
3. Bearing	14. Drive gear	20 & 20A. Transmission drive countershaft	32. Snap ring
4. Pto input shaft	15 & 15A. Bearings		33. Bearing
	16 & 16A. Thrust washer		34. Retainer
5. Bearing		21 & 21A. Bearing	35. Transmission input shaft
6. Range gear countershaft	17. Cluster gear & shaft assy.	22 & 22A. Snap ring	
7. Bearings	17A. Cluster gear (early models)	23. Pilot bearing	36. Bearing
8. Snap ring		24A. Snap rings	37. Pto counter-shaft
9. Low range gear	18. Needle bearing	25. Low range sliding gear	38. Gear
10. Spacer collar	18A. Needle bearings	26. Steel ball (2 used)	39. Snap ring
11. High range gear		27. High range sliding gear	40. Bearing
			41. Snap ring
			42. Coupling

shaft forward until front bearings are free of the housing, then remove bearings (15) and thrust washer (16) from front of countershaft. Tap countershaft rearward from housing while removing cluster gear and shaft assembly (17) and thrust washer (19) out top of housing.

On all models, remove range gear countershaft (6) and bearings (7). Remove snap ring (41) from pto countershaft rear bearing bore. Tap countershaft (37) forward until front bearing (36) is free of housing and remove the bearing. Remove countershaft, gear and rear bearing as an assembly from housing.

Inspect all parts for excessive wear or damage and renew if necessary.

To reassemble, reverse the disassembly procedure.

Fig. 193 — Rear view of early production range transmission housing showing bevel pinion shaft front bearing and support assembly (1).

NONSYNCHROMESH TRANSMISSION

(1300-1310-1500-1510-1710 Models)

The nonsynchromesh transmission used on these tractors consists of two sections. The forward section of transmission provides three forward speeds and one reverse. The rear section of transmission provides four range speeds. Together the main and range sections provide the tractor with twelve forward speeds and four reverse speeds.

LUBRICATION

All Models So Equipped

120. Transmission housing, center housing (differential and final drive), rear axles and hydraulic system share a common sump. The fluid should be drained and refilled with Ford 134 or equivalent fluid every 300 hours of operation. The oil drain plugs (Fig. 195) are located at lowest part of transmission housing, each final drive reduction compartment and front wheel drive housing if equipped with four wheel drive. Capacity is approximately 20 liters (21 U.S. quarts) for 1300 and 1500 models, 26 liters (27.5 U.S. quarts) for 1310 and 1510 models and 27 liters (28.5 U.S. quarts for 1710 models. Fluid level should be maintained between the full mark and lower end of dipstick which is located in shift cover. Fill opening is located either at transmission top cover or at top rear of hydraulic lift cover.

The hydraulic system filter is located at left side of engine by the hydraulic pump on 1300 and 1710 models, at right side of engine by hydraulic pump on 1310 and 1500 models and on left side of transmission housing on 1510 models. The filter element should be cleaned every 300 hours of operation on 1300, 1310 and 1500 models. Renew filter element if it is damaged or if it cannot be fully cleaned. On 1510 and 1710 models, spin-on type filter should be renewed every 300 hours of operation.

REMOVE AND REINSTALL

All Models So Equipped

121. To remove transmission, first remove drain plugs (Fig. 195) and drain oil from housings. Separate engine from clutch housing as outlined in paragraph 101. Remove fuel tank, steering gear and instrument panel. Disconnect clutch control rod from release shaft arm. Un-

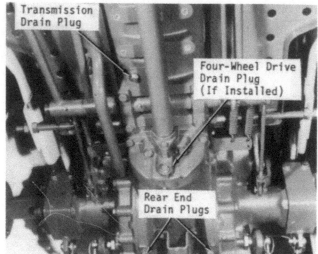

Fig. 195 — View showing location of oil drain plugs on 1300, 1310, 1500, 1510 and 1710 models.

bolt and remove clutch housing from transmission housing.

Disconnect foot throttle control rod and brake control rods. Remove both step plates. On models with front wheel drive, unbolt and remove front wheel drive gear case from rear axle center housing. On all models, remove brake cross shaft, brake pedals and clutch pedal as an assembly. Disconnect hydraulic suction line, pressure lines and return line. Remove shift detent plugs, springs and balls (49 – Fig. 196) from transmission shift cover. Remove shift cover and center housing front cover plate.

Support rear axle center housing at front and rear to prevent tipping. Support transmission housing with an overhead hoist. Remove nuts and bolts attaching transmission housing to center housing (note the two internal nuts at top of housing), then separate transmission from rear axle center housing.

When reinstalling transmission, use suitable liquid gasket maker on transmission housing mounting surfaces. Reinstall by reversing the removal procedure.

OVERHAUL

All Models So Equipped

122. To disassemble, remove retaining wires through shift forks and bosses. Drive roll pins from shift rails, then slide range shift rails (37 and 40 – Fig. 197) forward and remove shift forks (38 and

41) and bosses (39 and 42). Remove magnets (46) from main shift rails. Unbolt retainer (14) from rear of transmission, then withdraw range mainshaft (12) and bearing (13) rearward while removing gears (10 and 11). Slide main shift rails (43 and 45) out of transmission housing while withdrawing shifter forks (44 and 36) and bosses. Note that it may be necessary to rotate main shift rails to move interlock pins (52 – Fig. 196 and 197) out of notches in rails if rails do not slide out easily.

Remove pto input shaft (61 – Fig. 198) if so equipped.

Remove bearing retaining ring (1 – Fig. 197 or 198), then withdraw transmission input shaft (3) and bearing (2) from front of housing. Move transmission mainshaft (9) forward until bearing (8) is free from housing bore, then withdraw mainshaft and gears (5

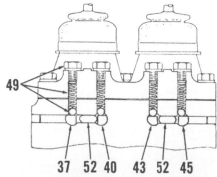

Fig. 196 — Cross-section of shift cover and housing showing detent assemblies (49), interlock plungers (52) and related parts. Refer to Fig. 197 for legend.

and 6) as an assembly through top opening of housing.

Remove snap ring (30) retaining rear countershaft rear bearing (29). Pry rear countershaft (28) and rear bearing rearward from housing while removing gears (24, 25 and 26) and spacer (57) from front of countershaft.

On models equipped with engine driven live pto, use a brass drift to drive pto drive shaft (66–Fig. 198) and front bearing (69) forward out of housing. Lift pto drive gear (67) out top opening and remove rear bearing (65) from housing bore.

On all models, remove snap ring (15–Fig. 197 or 198) retaining front countershaft front bearing (16). Tap countershaft (22–Fig. 197 or 64–Fig. 198) with front bearing forward out of housing. Withdraw countershaft gears (17, 19 and 21) and spacers (18 and 20)

out top of housing. Remove bearings (55 and 56) from transmission center web.

Remove reverse idler shaft retaining bolt (35) from housing. Drive roll pins (32P) out of shaft, then pull shaft rearward while removing thrust washers (31), idler gear (34) and bearing (33).

Inspect all parts for excessive wear or damage and renew if necessary. Backlash between gears should be 0.20 mm (0.008 inch) with a wear limit of 0.60 mm (0.024 inch). Clearance (C–Fig. 199) groove of sliding gears (B) and pad of shift forks (A) should be 0.10-0.30 mm (0.004-0.012 inch) with a wear limit of 1.0 mm (0.039 inch). Diametral clearance between shift rails and housing bores should be 0.03-0.10 mm (0.001-0.004 inch) with a wear limit of 0.30 mm (0.012 inch).

Lubricate all bearings with oil during assembly. When installing reverse idler,

install gear (34–Fig. 197 or 198) with longer hub side facing forward and notched surfaces of thrust washers (31) toward the gear. Install retaining bolt (35) with new seal washer making sure bolt engages hole in shaft (32).

Install front and rear countershafts making sure that gears and spacers are correctly positioned as shown in Fig. 197 or 198.

When installing mainshaft (9), be sure that bearing locking pin (8A) is positioned in housing with notched side facing forward and that mainshaft rear bearing seats against the notch. Make certain mainshaft sliding gears (5 and 6) are positioned with shift fork grooves facing rearward.

Install main shift forks, rails and magnets making sure that interlock pins (52) are correctly positioned in housing bores. Install roll pins in forks, bosses

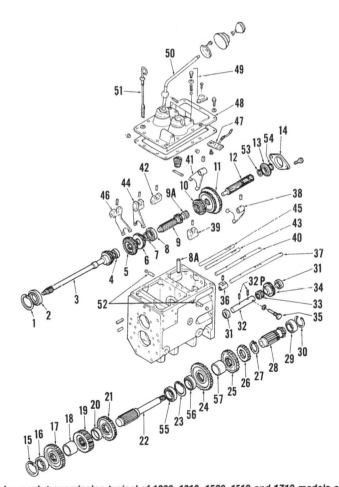

Fig. 197—Exploded view of nonsynchromesh transmission typical of 1300, 1310, 1500, 1510 and 1710 models equipped with transmission driven pto.

1. Snap ring	9A. Needle bearing
2. Bearing	10. Range sliding gear (3rd/4th)
3. Input shaft	11. Range sliding gear (1st/2nd)
4. Bearing	12. Range mainshaft
5. Sliding gear (1st/3rd)	13. Bearing
6. Sliding gear (2nd/reverse)	14. Retainer
7. Snap ring	15. Snap ring
8. Bearing	16. Bearing
8A. Retainer pin	17. Counter gear
9. Mainshaft (main transmission)	18. Spacer

19. Gear (1st)	29. Bearing
20. Spacer	30. Snap ring
21. Gear (2nd)	31. Thrust washers
22. Countershaft (main transmission)	32. Reverse shaft
23. Snap ring	32P. Pins
24. Counter gear	33. Needle bearing
25. Range gear (3rd)	34. Reverse idler gear
26. Range gear (2nd)	35. Retaining screw
27. Snap ring	36. Magnet (late models)
28. Range countershaft	37. Shift rod (range 1st/2nd)

38. Fork (range 1st/2nd)	47. Switch
39. Boss (range 1st/2nd)	48. Top cover
40. Shift rod (range 3rd/4th)	49. Detent plug, spring & ball (4 used)
41. Fork (range 3rd/4th)	50. Shift lever
42. Boss (range 3rd/4th)	51. Dipstick
43. Shift rod (1st/3rd)	52. Interlock pins (plungers)
44. Fork (main 1st/3rd)	53. Snap ring
45. Shift rod (2nd/Reverse)	54. Snap ring
46. Fork (main 2nd/reverse)	55. Bearing
	56. Bearing
	57. Spacer

and magnets, then safety wire in place. Be sure safety wires will not interfere with shifting.

Install rear mainshaft (12) from the rear while positioning range sliding gears (10 and 11) in housing as shown in Fig. 197 or 198. Install range shift forks, bosses and shift rails. Drive in roll pins and safety wire in place. Be sure shift interlock pins (52) are correctly positioned between shift rails.

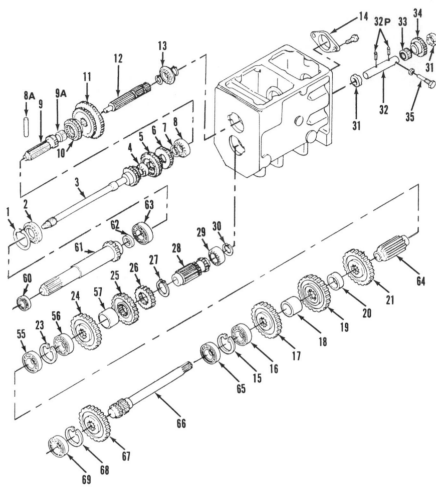

Fig. 198 — Exploded view of nonsynchromesh transmission typical of 1310, 1510, and 1710 models equipped with optional engine driven live pto. Refer to Fig. 197 for legend except for following parts.

60. Bearing
61. Pto input shaft
62. Oil seal
63. Bearing
64. Transmission countershaft
65. Bearing
66. Pto drive shaft
67. Drive gear
68. Snap ring
69. Bearing

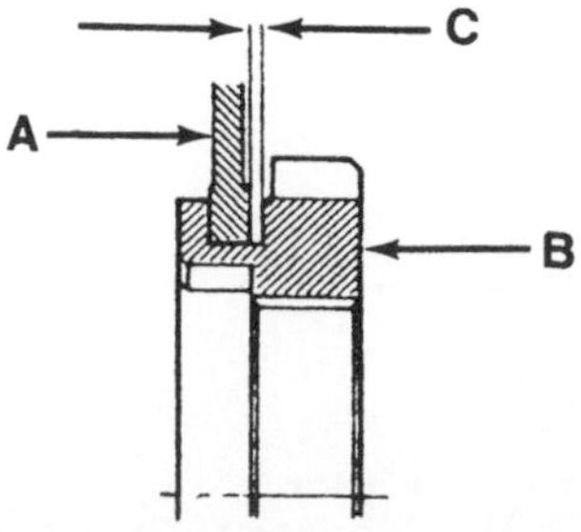

Fig. 199 — Clearance (C) between groove in transmission sliding gears (B) and shift fork (A) should not exceed 1.0 mm (0.039 inch).

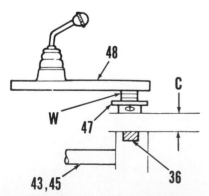

Fig. 200 — Clearance (C) between neutral safety start switch (47) and actuating magnets (36) on main shift rails (43 & 45) should be 3.5-4.0 mm (0.138-0.157 inch). Refer to text.

NONSYNCHROMESH TRANSMISSION

(1700 and 1900 Models With Single Plate Clutch)

The transmission used on all 1700 models is shown in Fig. 202. The transmission used on 1900 models with single plate engine clutch is shown in Fig. 203. The forward section of both transmissions provides three forward speeds and one reverse. The rear section contains four range speeds. Together, the main and range sections provide the tractor with twelve forward speeds and four reverse speeds.

LUBRICATION

All Models So Equipped

123. Transmission, center housing (differential and final drive), rear axles and hydraulic system share a common sump. The drain plugs are located in bottom of transmission housing and in bottom of rear axle center housing and in drive housing of models with four wheel drive. Refer to Fig. 201. The fluid should be drained and refilled with Ford 134 or equivalent fluid every 300 hours of operation. Capacity is approximately 22 liters (23 U.S. quarts) for Model 1700 and 24 liters (25.4 U.S. quarts) for Model 1900.

The hydraulic oil filter is located on lower right side of rear axle center housing (Fig. 201). Fluid must be drained from center housing before removing the filter. Filter should be cleaned every 300 hours of operation. Renew the filter if it is damaged or if it cannot be cleaned satisfactorily.

REMOVE AND REINSTALL

All Models So Equipped

124. To remove transmission, first drain fluid from housings. Separate engine from clutch housing as outlined in paragraph 101. Remove fuel tank, steering gear and instrument panel. Disconnect clutch control rod from release shaft arm. Unbolt and remove clutch housing from transmission housing.

Disconnect foot throttle control rod and brake control rod. Remove both step plates. Remove brake cross shaft, brake pedals and clutch pedals as an assembly. Disconnect hydraulic suction line, pressure lines and return lines. Remove shift detent plugs, springs and balls (49 – Fig. 202 or Fig. 203) from shift cover. Unbolt and remove shift cover from transmission housing.

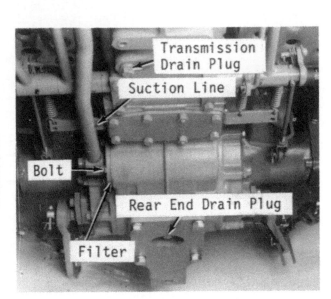

Fig. 201 – View showing location of oil drain plugs for transmission housing, rear axle center housing and hydraulic system on 1700 and 1900 models. Oil must be drained before removing hydraulic filter.

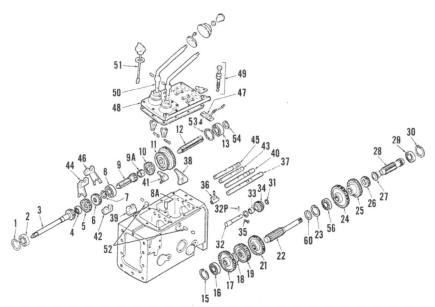

Fig. 202 – Exploded view of transmission used on 1700 models.

2. Bearing
3. Input shaft
4. Bearing
5. Sliding gear (1st/3rd)
6. Sliding gear (2nd/Reverse)
7. Snap ring
8. Bearing
8A. Retainer pin
9. Mainshaft (main transmission)
9A. Needle bearing
10. Range sliding gear (3rd/4th)
11. Range sliding gear (1st/2nd)
12. Range mainshaft
13. Bearing
16. Bearing
17. Counter gear
18. Spacer
19. Gear (1st)
21. Gear (2nd)

22. Countershaft (main transmission)
23. Snap ring
24. Counter gear
25. Range gear (3rd)
26. Range gear (2nd)
27. Snap ring
28. Range countershaft
29. Bearing
30. Snap ring
31. Thrust washer
32. Reverse shaft
32R. Thrust washer
33. Bearing
34. Reverse idler
35. Pin
36. Magnet (late models)
37. Shift rod (Range 1st/2nd)
38. Fork (Range 1st/2nd)
39. Boss (Range 1st/2nd)

40. Shift rod (Range 3rd/4th)
41. Fork (Range 3rd/4th)
42. Boss (Range 34rd/4th)
43. Shift rod (1st/3rd)
44. Fork (Main 1st/3rd)
45. Shift rod (2nd/Reverse)
46. Fork (Main 2nd/Reverse)
47. Switch
48. Top cover
49. Detent plug, spring and ball (4 used)
50. Shift lever
51. Dipstick
52. Interlock pins (plungers)
53. Snap ring
54. Snap ring
56. Bearing
58. Retainer
59. "O" ring
60. Thrust washer

Support rear axle center housing at front and rear to prevent tipping. Support transmission housing with an overhead hoist. Remove nuts and bolts attaching transmission housing to center housing, then slide transmission forward off the studs.

To reinstall, reverse the removal procedure. Apply liquid gasket maker to transmission housing mounting surfaces being careful not to get excess gasket maker inside the housings.

OVERHAUL

All Models So Equipped

125. To disassemble, remove retaining wires through shifter forks and bosses. Drive roll pins from shift rails, then slide range shift rails (37 and 40–Fig. 202 or Fig. 203) from housing while removing shift forks (38 and 41) and bosses (39 and 42). Move range mainshaft (12) rearward while withdrawing range sliding gears (10 and 11). Remove needle bearing (9A) from end of mainshaft. Slide main shift rails (43 and 45) out rear of housing while removing safety start switch magnets (36) and shift forks (44 and 46).

Remove snap ring (1–Fig. 203) on 1900 models retaining input shaft front bearing. On all models, withdraw input shaft (3–Fig. 202 or Fig. 203) from front of housing. Remove mainshaft rear bearing retaining pin (8A). Move mainshaft (9), bearing (8) and sliding gears (5 and 6) forward until bearing is free from housing bore, then withdraw as an assembly through top opening.

On Model 1700, tap main transmission countershaft (22–Fig. 202) forward to remove retainer (58), bearing (16) and countershaft from housing. Remove gears (17, 19 and 21), spacer (18) and thrust washer (60) through top opening.

On Model 1900, remove snap ring (15–Fig. 203) retaining countershaft front bearing. Tap main transmission countershaft (22) and front bearing (16) forward from housing. Remove gears (17, 19 and 21), spacer (18) and rear bearing (60) through top of housing.

On all models, remove snap ring (30–Fig. 202 or Fig. 203) retaining range countershaft rear bearing. Slide range countershaft (28) and rear bearing (29) rearward from housing while withdrawing gears (24, 25 and 26) out top of housing.

To remove reverse idler assembly on 1700 models, slide idler shaft (32–Fig. 202) forward out of housing bores while removing thrust washers and idler gear (34) with needle bearing (33) out top opening.

To remove reverse idler assembly on 1900 models, remove retaining bolt (35–Fig. 203) and drive retaining pin (32P) out of idler shaft (32). Push shaft forward out of housing bore and remove thrust washers and idler gear (34) with needle bearing (33) from housing.

Inspect all parts for excessive wear or damage and renew if necessary.

To reassemble, reverse the disassembly procedure while noting the following special instructions.

On 1700 models, install reverse idler gear with longer shoulder side facing rearward. On 1900 models, install reverse idler gear with longer shoulder side toward the front.

When installing transmission mainshaft (9), be sure that notch of retaining pin (8A) correctly engages bearing (8). Be sure mainshaft sliding gears (5 and 6) are positioned with grooves for shift forks facing rearward.

Be sure that interlock pins (52) are correctly positioned in housing bores between the shift rails. Notches in shift rails for detent balls (39) should face upward. Install roll pins in shift forks and safety wire in place. Make certain safety wires will not interfere with shifting.

Fig. 205 – View of oil drain plugs on 1900, 1910 and 2110 models. Hydraulic filter is located in right side of rear axle center housing on 1900 models.

Fig. 206 – View showing location of hydraulic oil filter on 1910 and 2110 models.

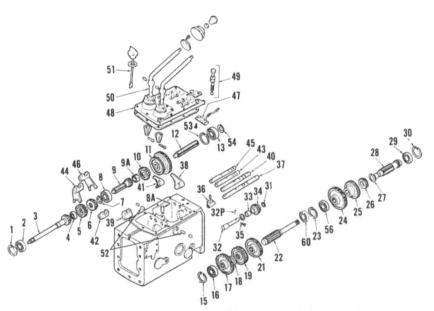

Fig. 203 — Exploded view of transmission used on 1900 models equipped with single plate clutch (transmission driven pto). Refer to Fig. 202 for legend except for the following.

1. Snap ring
15. Snap ring
32P. Pin
35. Retainer screw
60. Bearing

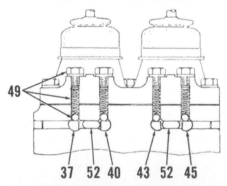

Fig. 207 – Cross section of shift cover and transmission housing showing detent assemblies (49), interlock plungers (52) and shift rails (37, 40, 43 and 45).

NONSYNCHROMESH TRANSMISSION

(1900-1910-2110 Models With Dual Plate Clutch)

The nonsynchromesh transmission used on 1900 and 1910 models equipped with dual plate clutch (engine driven pto) is shown in Fig. 209. The nonsynchromesh transmission used on 2110 models is similar and is shown in Fig. 210 and 211. The forward section of transmission provides three forward speeds and one reverse. The rear section provides four range speeds. Together, the main and range sections provide the tractor with twelve forward speeds and four reverse speeds.

LUBRICATION

All Models So Equipped

126. Transmission, center housing (differential and final drive), rear axle and hydraulic system share a common sump. The oil drain plugs are located in bottom of transmission housing, in bottom of rear axle center housing and in front wheel drive gear case if so equipped. Refer to Fig. 205. The fluid should be drained and refilled with new Ford 134 or equivalent fluid every 300 hours of operation. Capacity is approximately 24 liters (25.4 U.S. quarts) for 1900 models, 28 liters (29.5 U.S. quarts) for 1910 models and 32.2 liters (34 U.S. quarts) for 2110 models.

On 1900 models, the hydraulic system filter is located in right side of rear axle center housing (Fig. 205). Fluid must be drained before removing filter. The filter element should be cleaned after every 300 hours of operation. Renew filter if it is damaged or if it cannot be cleaned satisfactorily.

On 1910 and 2110 models, a spin-on type hydraulic filter is located at right front corner of engine next to hydraulic pump (Fig. 206). Filter should be re-

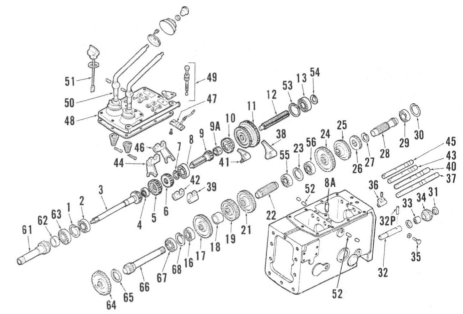

Fig. 209 — Exploded view of nonsynchromesh transmission used on 1900 and 1910 models which are equipped with dual plate engine clutch (live pto).

1. Snap ring	16. Bearing	33. Bearing
2. Bearing	17. Counter gear	34. Reverse idler
3. Input shaft	18. Spacer	35. Retaining screw
4. Bearing	19. Gear (1st)	36. Magnet (late models)
5. Sliding gear (1st/3rd)	21. Gear (2nd)	37. Shift rod (range 1st/2nd)
6. Sliding gear (2nd/Reverse)	22. Countershaft (main transmission)	38. Fork (range 1st/2nd)
7. Snap ring	23. Snap ring	39. Boss (range 1st/2nd)
8. Bearing	24. Counter gear	40. Shift rod (range 3rd/4th)
8A. Bearing retainer pin	25. Range gear (3rd)	41. Fork (range 3rd/4th)
9. Mainshaft (main transmission)	26. Range gear (2nd)	42. Fork (range 3rd/4th)
9A. Bearing	27. Snap ring	43. Shift rod (1st/3rd)
10. Range sliding gear (3rd/4th)	28. Range countershaft	44. Fork (main 1st/3rd)
11. Range sliding gear (1st/2nd)	29. Bearing	45. Shift rod (2nd/reverse)
12. Range mainshaft	30. Snap ring	46. Fork (main 2nd/reverse)
13. Bearing	31. Thrust washer	47. Switch
	32. Reverse shaft	
	32P. Pin	
48. Top cover		
49. Detent plug, spring and ball (4 used)		
50. Shift lever		
51. Dipstick		
52. Interlock pins (plungers)		
53. Snap ring		
54. Snap ring		
55. Bearing		
56. Bearing		
61. Pto input shaft		
62. Oil seal		
63. Bearing		
64. Pto driven gear		
65. Snap ring		
66. Pto front shaft		
67. Bearing		
68. Snap ring		

newed after every 300 hours of operation. Prior to removing filter, remove plug from suction tube banjo bolt, set throttle in "stop" position and crank engine for about 10 seconds to drain oil from filter. Install new filter and reinstall plug in banjo bolt.

Fig. 208 — View showing removal of detent springs from cover bores. Springs may be lost or damaged if not removed before lifting cover from transmission.

REMOVE AND REINSTALL

All Models So Equipped

127. To remove transmission, first drain oil from transmission housing and rear axle center housing. Separate engine from clutch housing as outlined in paragraph 101. Remove fuel tank, instrument panel and steering gear assembly. Disconnect clutch control rod, brake control rods and foot throttle control rod. Remove both foot step plates. Remove brake cross shaft and pedals as an assembly. Unbolt and remove clutch housing from transmission housing.

Remove detent plugs, springs and balls (49 – Fig. 207) from transmission shift cover. Remove the transmission shift cover and remove cover plate or shift cover from front of axle center housing. Support rear axle center housing at the front and rear to prevent tipping. Support transmission housing with

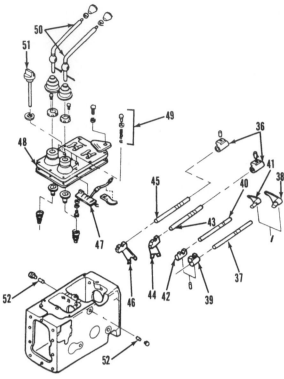

from housing bore and remove main-shaft and sliding gears (5 and 6) from housing.

On 2110 models, straighten tabs of lockwasher (71–Fig. 211) and loosen locknuts (70) on rear countershaft (28). On all models, remove snap ring (30–Fig. 209 or 211) retaining counter-shaft rear bearing. Remove counter-shaft and rear bearing rearward from housing while withdrawing gears (24, 25 and 26) out top opening. Remove snap ring (68) retaining front countershaft bearing. Drive front countershaft (22) and bearing (16) forward out of housing and remove gears (17, 19 and 21) and spacer (18) out top of housing.

To remove reverse idler assembly, re-move retaining bolt (35) from idler shaft

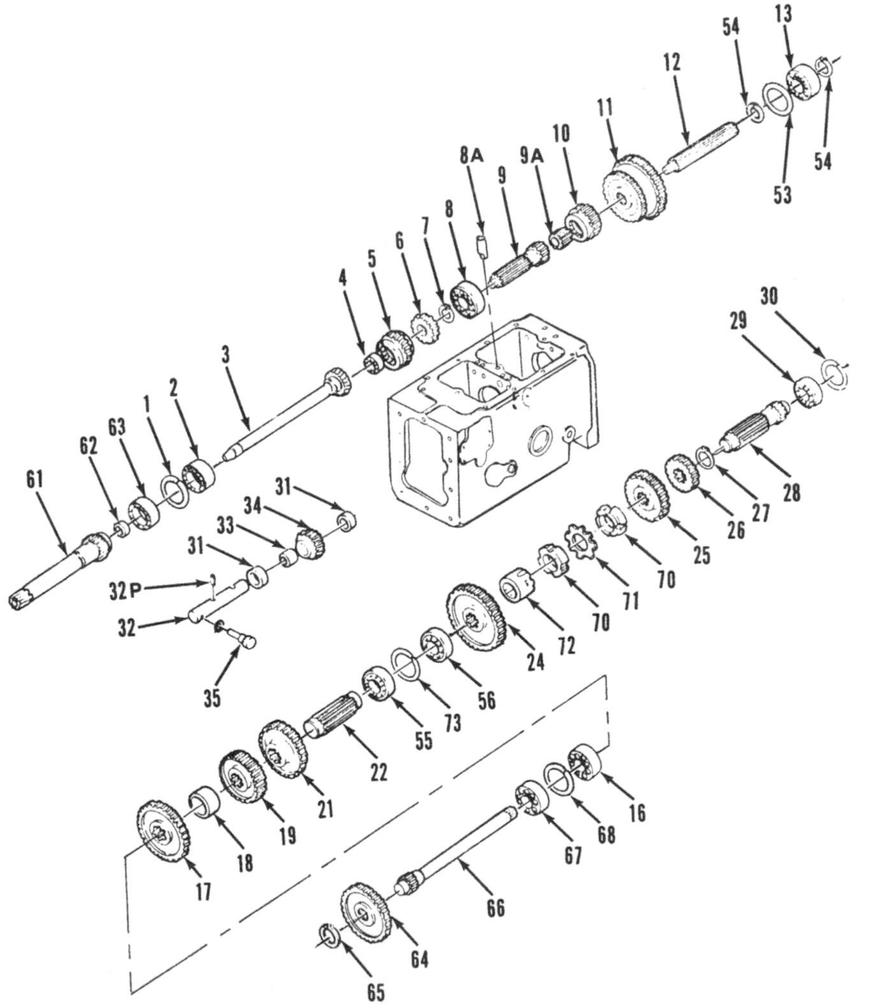

Fig. 210 – Exploded view of shift control linkage used on 2110 models with nonsyn-chromesh transmission. Refer to Fig. 209 for legend.

an overhead hoist. Remove nuts and bolts attaching transmission to rear axle center housing, then pry transmission away from center housing.

To reinstall, reverse the removal pro-cedure. Apply liquid gasket maker to transmission mounting surfaces being careful not to get excess sealer inside the housings.

OVERHAUL

All Models So Equipped

128. Refer to appropriate Figs. 209, 210 and 211 for exploded view of trans-mission and shift linkage. To dis-assemble, remove safety wires and drive roll pins out of shift rails. Slide range gear shift rails forward out of housing while removing shift forks (38 and 41) and bosses (39 and 42). Withdraw range mainshaft (12) and bearing (13) from rear of housing and remove range sliding gears (10 and 11) out top of hous-ing. Slide main gear shifter rails from housing while removing shift forks (44 and 46) and neutral start switch magnets (36).

Tap pto drive shaft (66) with gear (64) and bearing (67) forward out of housing. Withdraw pto input shaft (61) and bear-ing (63) from front of housing. Remove snap ring (1), then pull transmission in-put shaft (3) with bearing (2) out front of housing. Remove pin (8A) retaining mainshaft rear bearing (8). Move main-shaft (9) forward until bearing (8) is free

Fig. 211 – Exploded view of nonsynchromesh transmission used on some 2110 models.

1. Snap ring	9A. Bearing	25. Gear (41T)	55. Bearing
2. Bearing	10. Range sliding gear	26. Gear (28T)	56. Bearing
3. Transmission input	(3rd/4th)	27. Snap ring	61. Pto input shaft
shaft	11. Range sliding gear	28. Range countershaft	62. Oil seal
4. Bearing	(1st/2nd)	29. Bearing	63. Bearing
5. Sliding gear	12. Range mainshaft	30. Snap ring	64. Pto driven gear
(1st/3rd)	13. Bearing	31. Thrust washers	65. Snap ring
6. Sliding gear (2nd/	16. Bearing	32. Idler shaft	66. Pto front shaft
reverse)	17. Gear (48T)	32P. Pin	67. Bearing
7. Snap ring	18. Spacer	33. Bearing	68. Snap ring
8. Bearing	19. Gear (41T)	34. Reverse idler gear	70. Adjusting nuts
8A. Bearing retaining	21. Gear (45T)	35. Retaining bolt	71. Lockwasher
pin	22. Main countershaft	53. Snap ring	72. Collar
9. Mainshaft (main	24. Gear (52T)	54. Snap rings	73. Spacer
transmission)			

(32). Drive out roll pins (32 P), then slide shaft forward while removing thrust washers (31), idler gear (34) and bearing (33).

Inspect all parts for excessive wear or damage and renew if necessary. Backlash between gears should not exceed 0.60 mm (0.024 inch). Side clearance (C – Fig. 199) between sliding gear shifter grooves and shift forks should be 0.10-0.30 mm (0.004-0.012 inch) with a wear limit of 1.0 mm (0.039 inch). Diametral clearance between shift rails and housing bores should be 0.03-0.10 mm (0.001-0.004 inch) with a wear limit of 0.30 mm (0.012 inch).

To reassemble transmission, reverse the removal procedure while noting the following special instructions.

When installing reverse idler gear assembly, be sure that grooved face of thrust washers (31) is positioned toward idler gear (34).

Install front countershaft (22) assembly first, then install rear countershaft (28) assembly. On Model 2110, turn adjusting nuts (70 – Fig. 211) on rear countershaft until all gear end play is removed. Bend tabs of lockwasher (71) to secure the nuts.

When installing front mainshaft (9 – Fig. 209 and 211) assembly, be sure

that retaining pin (8A) is positioned with notched side toward the front and the bearing (8) is seated against the notch. Install main gear shift rails and forks, rear mainshaft (12) assembly and range gear shift rails and forks. Be sure that interlock pins (52 – Fig. 207) are properly positioned between shift rails and that notches in shift rails face the pins. Only one shift rail may be engaged at a time in main transmission and range gear compartments.

When installing pto input shaft (61 – Fig. 209 and 211) over transmission input shaft (3), be careful not to damage oil seal (62).

SYNCHROMESH TRANSMISSION

(1310-1510-1710-1910-2110 Models)

An optional synchromesh transmission is available for 1310, 1510, 1710, 1910 and 2110 models. The forward section of transmission contains main transmission gears consisting of three forward and one reverse speed gears which are in constant mesh and synchronized. These gears are controlled by shift lever mounted on steering column. The rear compartment contains four range speed gears which are not synchronized. The sliding range gears are controlled by a shift lever located on transmission cover. Together, the main and range sections provide twelve forward speeds and four reverse speeds.

LUBRICATION

All Models So Equipped

130. The transmission housing, center housing (differential and final drives), rear axles and hydraulic system share a common oil reservoir. The oil drain plugs are located in bottom of transmission housing, bottom of rear axle center housing and in drive housing for front wheel drive if so equipped. The fluid should be drained and refilled with Ford 134 or equivalent fluid every 300 hours of operation. Capacity is approximately 26 liters (27.5 U.S. quarts) for 1310 and 1510 models, 27 liters (28.5 U.S. quarts) for 1710 and 1910 models and 32 liters (34 U.S. quarts) for 2110 models. Fluid level should be maintained between the full mark and lower end of dipstick which is located in transmission cover.

The hydraulic system filter is located on right side of engine by the hydraulic

pump on 1310, 1910 and 2110 models, on left side of transmission housing on 1510 models and on left side of engine by the hydraulic pump on 1710 models. On 1310 models, the filter should be cleaned every 300 hours of operation. Filter must be renewed if it is damaged or if it cannot be cleaned satisfactorily. On 1510, 1710, 1910 and 2110 models, spin-on type hydraulic filter should be renewed every 300 hours of operation.

REMOVE AND REINSTALL

All Models So Equipped

131. To remove transmission, first drain oil from housings. Separate engine from clutch housing as outlined in paragraph 101. Disconnect clutch control rod from release shaft arm. Disconnect foot throttle control rod and brake control rods. Remove brake cross shaft and pedals as an assembly. Remove both foot step plates. Remove console cover (12 – Fig. 215) and disconnect main transmission shift rod (10) at the lower U-joint (8). Remove fuel tank, steering gear and instrument panel with shifter controls. Unbolt and remove clutch housing from transmission housing.

Disconnect hydraulic lines as necessary. Remove shift detent cover plate (2 – Fig. 214), springs (1) and balls. Remove shift cover from transmission housing. Remove cover plate or shift cover from front of rear axle center housing.

Support rear axle center housing at front and rear to prevent tipping. Support transmission housing with an overhead hoist. Remove nuts and bolts (not-

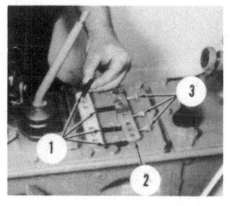

Fig. 214 — Shift detent springs (2) and balls should be removed prior to removing shift cover to avoid loss or damage of springs.

ing two internal nuts at top of housing) attaching transmission housing to axle center housing, then separate transmission from center housing.

When reinstalling transmission, use suitable liquid gasket maker on transmission housing mounting surfaces. Be careful not to get excess gasket maker inside the housings.

OVERHAUL

Models 1310-1510-1710

132. To disassemble transmission, remove safety wires from shift forks and drive roll pins from shift forks (17 and 20 – Fig. 215), bosses (16) and neutral start switch magnets (19). Remove the magnets from main shift rails (18). Slide main gear shift rails out rear of housing while withdrawing upper shift fork. The lower shift fork can

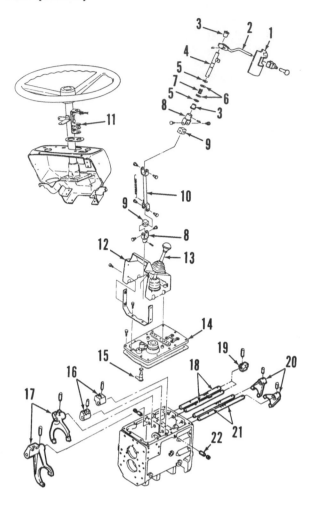

Fig. 215—Exploded view of shift linkage used on 1310, 1510, 1710, 1910 and 2110 models equipped with synchromesh transmission.

1. Cover
2. Shift lever (main transmission)
3. Bushings
4. Shifter rod (upper)
5. Snap rings
6. Washers
7. Spring
8. U-joint
9. Block
10. Shifter rod (lower)
11. Support bracket
12. Console cover
13. Shift lever (range)
14. Shift cover
15. Shifter arm
16. Bosses
17. Main gear shift forks
18. Main gear shift rails
19. Magnet (2 used)
20. Range gear shift forks
21. Range gear shift rails
22. Interlock pin

be removed after transmission mainshaft is removed. Slide range shift rails (21) rearward from housing while removing shift bosses (16) and forks (20).

Remove cap screws from rear bearing retainer (31–Fig. 216), then withdraw range mainshaft (30) with bearing (29) rearward from housing. Remove range sliding gears (26 and 27).

Remove snap ring (41) retaining range countershaft rear bearing (42). Slide countershaft (43) and rear bearing rearward from housing while removing counter gears (48, 46 and 45) and spacer (47) out top of housing.

Remove input shaft retainer and housing (32). Withdraw pto input shaft (2) and pto driven gear (61) and shaft (62) from front of housing.

Remove snap ring (23) retaining mainshaft rear bearing, then withdraw mainshaft (24) and bearing (22) rearward from housing. Remove synchronizer assembly, constant mesh gears (18 and 20) and spacer (19) through top opening as shaft is withdrawn. Be careful not to lose steel ball (16), if used, that retains coupling (15) to mainshaft.

Remove rear bearing (49) and snap ring (50) from housing bore, then tap

main countershaft (59) rearward. Remove gears (58, 56 and 54), spacer (57), coupling (55), synchronizer assembly and thrust washer (51) as countershaft is withdrawn from housing. Be careful not to lose steel balls (16) if used.

Move transmission input shaft (8) and bearing rearward and upward through top opening of transmission housing.

To remove reverse idler assembly, remove retaining bolt (37) and drive roll pins out of idler shaft (36). Slide shaft out rear of housing as thrust washers (38), idler gear (40) and bearings (39) are removed through top opening.

Inspect all parts for excessive wear or damage. Check synchronizer cone wear by measuring clearance (S–Fig. 217) between teeth of synchronizer ring (1) and gear (2). Renew parts as necessary if clearance is less than 0.5 mm (0.020 inch). Renew synchronizer ring and/or gear if teeth are worn or broken. Clearance (C–Fig. 218) between shifter forks (3) and grooves of synchronizer collars (4) and range sliding gears should be 0.10-0.30 mm (0.004-0.012 inch) with a wear limit of 1.0 mm (0.039 inch).

To reassemble transmission, reverse the removal procedure while noting the following special instructions.

When installing reverse idler assembly, be sure that hub side of idler gear (40–Fig. 216) faces rearward. Position grooved face of thrust washers (38) toward idler gear.

Install transmission input shaft (8) with bearing (6). Insert main countershaft (59) from the rear while assembling thrust washer (51), synchronizer assembly, gears (56 and 58) and spacer (57) onto shaft.

NOTE: Make sure steel balls (16), if used, that retain synchronizer collars (51 and 55) in a fixed position on countershaft are in place during assembly.

Install snap ring (50) to retain countershaft rear bearing, then use a dial indicator to measure countershaft end play. Maximum allowable end play is 0.20 mm (0.008 inch). Install shims (64), if necessary, in front of countershaft front bearing (60) to obtain desired end play.

Position lower shifter fork in groove of lower synchronizer sliding collar. Insert transmission mainshaft (24) from the rear while installing gears (18 and 20), spacer (19) and synchronizer assembly. Install snap-ring (23) retaining

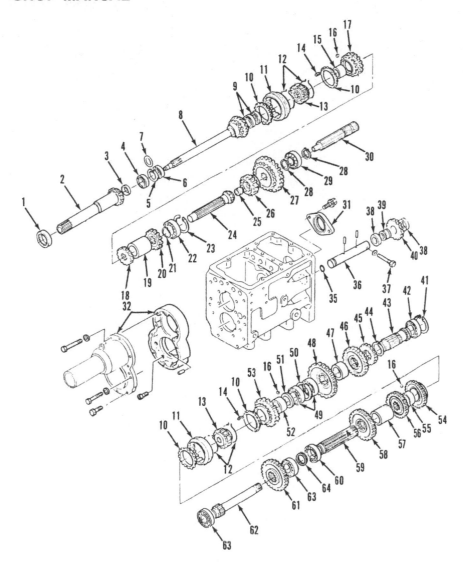

Fig. 216—Exploded view of optional synchromesh transmission used on some 1310, 1510, and 1710 models. Steel ball (16) is not used on early production models.

1. Bearing
2. Pto drive input shaft
3. Oil seal
4. Bearing
5. Snap ring
6. Bearing
7. Shim
8. Transmission input shaft
9. Bearing
10. Synchronizer rings
11. Sliding collar
12. Springs
13. Synchronizer hub
14. Key (3 used)
15. Coupling
16. Steel ball
17. Synchronizer gear (26T)
18. Constant mesh gear (20T)
19. Spacer
20. Constant mesh gear (22T)
21. Snap ring
22. Bearing
23. Snap ring
24. Mainshaft
25. Bearing
26. Range sliding gear (3rd/4th)
27. Range sliding gear (1st/2nd)
28. Snap rings
29. Bearing
30. Range mainshaft
31. Bearing retainer

32. Input shaft retainer housing
35. "O" ring
36. Idler shaft
37. Retainer bolt
38. Thrust washers
39. Bearing
40. Reverse idler gear
41. Snap ring
42. Bearing
43. Range gear countershaft
44. Snap ring
45. Counter gear (28T)
46. Counter gear (41T)
47. Spacer
48. Counter gear (51T)
49. Bearings
50. Snap ring
51. Thrust washer
52. Spacer
53. Synchronizer gear (33T)
54. Synchronizer gear (32T)
55. Coupling
56. Gear (30T)
57. Spacer
58. Gear (37T)
59. Main transmission countershaft
60. Bearing
61. Gear (37T)
62. Pto front driven shaft
63. Bearing
64. Shim

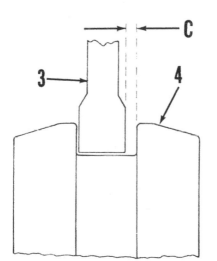

Fig. 218 — Side clearance (C) between shift forks (3) and grooves of sliding collars (4) and range sliding gears should not exceed 1.0 mm (0.039 inch).

rear bearing, then use a dial indicator to measure mainshaft end play. Maximum allowable end play is 0.20 mm (0.008 inch). To adjust end play, install shims (7) between input shaft front bearing (6) and snap ring (5).

Install range countershaft (43) assembly, range mainshaft (30) assembly and shift rails, forks and bosses. Be sure to safety wire shift rail roll pins.

Models 1910-2110

133. Refer to appropriate Fig. 220 or Fig. 221 for exploded view of synchromesh transmission. To disassemble transmission, first withdraw range gear countershaft (32) rearward while removing sliding gears (27 and 28) through top of housing.

Drive roll pins from shifter forks (17 and 20 – Fig. 215), bosses (16) and safety start switch magnets (19). Remove magnets from main shift rails (18), then slide shift rails out rear of housing and withdraw upper shift fork. The lower

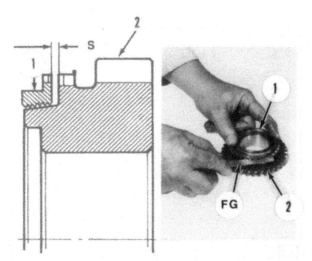

Fig. 217—Measure clearance (S) between teeth of synchronizer ring (1) and gear (2) using a feeler gage (FG) to check synchronizer cone wear.

shift fork can be removed after transmission mainshaft is removed. Slide range shift rails (21) rearward from housing while removing shift bosses (16) and forks (20).

Remove pto drive input shaft (3 – Fig. 220 or 221) with bearing (5) and oil seal (4). On Model 2110, remove coupling (44 – Fig. 221) from rear of pto driven shaft (75). Withdraw pto shaft and driven gear (72) forward from housing. On Model 1910, remove pto driven gear (72 – Fig. 220) and bearing (71) from front of housing. The pto driven shaft is removed with the rear axle center housing when transmission is separated from center housing.

On Model 2110, straighten tabs of lockwasher (65 – Fig. 221) and loosen the nuts (64). On all models, remove snap ring (45 – Fig. 220 or 221) retaining countershaft rear bearing (46). Slide range countershaft (48) and rear bearing rearward from housing. Remove counter gears (52, 51 and 50 out top of housing.

Remove snap ring (24) at rear of mainshaft rear bearing (23). Withdraw mainshaft (25) and rear bearing out rear of housing while removing synchronizer assembly (11 through 19) and gears (20 and 21). Remove lower shift fork.

Remove retaining bolt (41) and drive roll pins (37) out of reverse idler shaft (36). Slide shaft out of housing while withdrawing thrust washers (38), reverse idler gear (40) and bearing (39).

On Model 1910, straighten tab on lockwasher (65 – Fig. 220) and loosen nuts (64). On all models, remove snap ring (70 – Fig. 220 or 221) retaining countershaft front bearing (69). Drive countershaft (68) forward from housing while removing gears (63 and 67) and synchronizer assembly out top of housing.

Tap transmission input shaft (9) and bearing (8) inward and remove through top opening of housing.

Inspect all parts for excessive wear or damage and renew if necessary. Maximum allowable backlash between gears is 0.30 mm (0.012 inch). Check synchronizer cone wear by measuring clearance (S – Fig. 217) between synchronizer ring (1) and gear (2) using a feeler gage. Renew parts as necessary if clearance is less than 0.5 mm (0.020 inch). Renew ring and/or gear if teeth are worn or broken. Side clearance (C – Fig. 218) between shifter forks (3) and synchronizer sliding collars (4) or range sliding gears should not exceed 1.0 mm (0.039 inch).

To reassemble transmission, first install transmission input shaft (9 – Fig. 220 or 221) and bearings (8 and 10).

Insert main countershaft (68) from the front through the counter gears, synchronizer assembly, thrust washer and

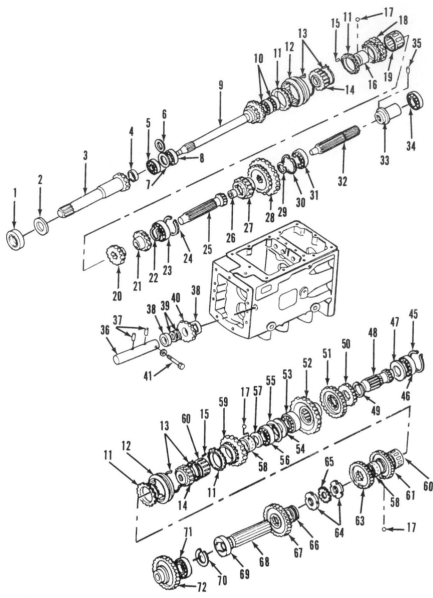

Fig. 220 – Exploded view of optional synchromesh transmission used on Model 1910.

1. Bearing	18. Synchronizer gear	34. Bearing
2. Shim	19. Bearing	35. Pin
3. Pto input shaft	20. Gear	36. Idler shaft
4. Oil seal	21. Gear	37. Pins
5. Bearing	22. Snap ring	38. Thrust washers
6. Shim	23. Bearing	39. Bearings
7. Snap ring	24. Snap ring	40. Reverse idler gear
8. Bearing	25. Transmission	45. Snap ring
9. Transmission input	mainshaft	46. Bearing
shaft	26. Bearing	47. Shim
10. Bearings	27. Range sliding gear	48. Range countershaft
11. Synchronizer rings	28. Range sliding gear	49. Snap ring
12. Sliding collar	29. Snap ring	50. Gear
13. Springs	30. Snap ring	51. Gear
14. Hub	31. Bearing	52. Gear
15. Key	32. Range mainshaft	53. Bearing
16. Collar	33. Coupling	54. Shim
17. Steel balls		

55. Snap ring	
56. Bearing	
57. Thrust washer	
58. Collars	
59. Synchronizer gear	
60. Bearings	
61. Synchronizer gear	
63. Gear	
64. Nuts	
65. Lockwasher	
66. Collar	
67. Gear	
68. Transmission	
countershaft	
69. Bearing	
70. Snap ring	
71. Bearing	
72. Pto drive gear	

rear bearing. Be sure that grooved face of thrust washer (57) faces the synchronizer gear (59) and that steel balls (17) are positioned in grooves in collars (58). On Model 1910, adjust nuts (64 – Fig. 220) to remove all end play from counter gears. Bend tabs of lockwasher (65) into slots in nuts to secure the nuts.

On all models, position lower shift fork

in groove of synchronizer sliding collar (12 – Fig. 220 or 221). Install reverse idler gear assembly making sure that grooved side of thrust washers (38) faces the idler gear (40). Install transmission mainshaft (25) from the rear while assembling gears (20 and 21) and synchronizer assembly on the shaft. Check mainshaft end play using a dial indicator. Maximum allowable end play is

0.20 mm (0.008 inch). Install shims (6), if necessary, in front of input shaft bearing to adjust end play.

On Model 1910, install two 0.20 mm (0.008 inch) shims (47–Fig. 220) and rear bearing (46) on range countershaft (48). Install front bearing (53) in housing bore, then insert countershaft (without gears) and secure with rear snap ring (45). Use a dial indicator to measure countershaft end play. Install shims (54) in front of front bearing to adjust end play as close to zero as possible. Max-

imum allowable end play is 0.20 mm (0.008 inch). After backlash is correctly set, withdraw countershaft and install gears (50, 51 and 52).

On Model 2110, install countershaft front bearing (53–Fig. 221) in housing bore. Insert range countershaft (48) from the rear while assembling gears (50, 51 and 52) and collar (66) with nuts (64) and lockwasher (65) on the shaft. Install rear snap ring (45), then adjust nuts (64) to remove all end play from counter gears. Bend tabs of lockwasher into

slots of nuts to lock the nuts in place.

On Model 1910, install coupling (33–Fig. 220) and retaining pin (35) onto rear of range mainshaft (32). Install bearing (31) onto front of coupling. Insert shaft and coupling from the rear while assembling sliding gears (27 and 28) on the shaft. On Model 2110, insert range mainshaft (32–Fig. 221) from the rear while installing sliding gears (27 and 28) on the shaft.

Complete assembly by reversing disassembly procedure.

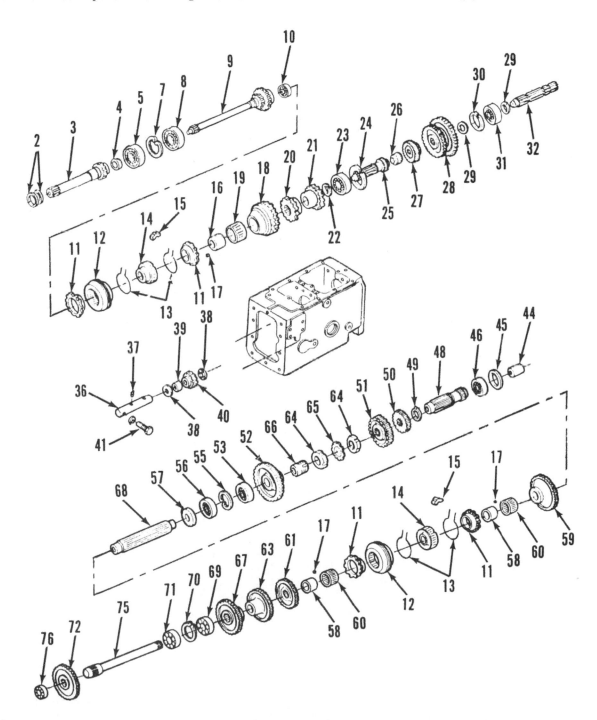

Fig. 221 – Exploded view of optional synchromesh transmission used on Model 2110. Refer to Fig. 220 for legend except for coupling (44), pto drive shaft (75) and bearing (76).

AXLE CENTER HOUSING AND DIFFERENTIAL

SPLIT TRANSMISSION FROM CENTER HOUSING

Models 1100-1110-1200-1210

135. Drain oil from transmission and center housing. Remove hydraulic lift arm housing as outlined in paragraph 180. Remove lower lift links and mounting bracket. Disconnect brake control rods. Remove transmission top cover. Disconnect wiring to rear of tractor. Suitably support rear axle center housing and transmission housing. Remove nuts and bolts attaching center housing to transmission housing, then move center housing rearward from transmission.

To reinstall, reverse the removal procedure. Be sure that shafts are correctly meshed before tightening nuts and bolts which attach center housing to transmission housing.

Models 1300-1310-1500-1510-1700-1710-1900-1910-2110

136. Drain oil from transmission and center housing. Disconnect clutch control rod, brake control rods and foot throttle control rod. Remove both step plates. On four wheel drive models, unbolt and remove front wheel drive gear case. On all models, remove brake cross shaft, brake pedals and clutch pedal as an assembly. Disconnect electrical wiring and hydraulic lines as necessary. On models with synchromesh transmission, remove console cover and disconnect lower shift control rod fork. On all models, remove detent springs and balls from transmission shift cover. Unbolt and remove shift cover. Remove top cover or shifter cover from center housing. Remove hydraulic lift cover as outlined in paragraph 180.

Install wedges between front axle and side rails to prevent tipping. Support rear axle center housing and transmission housing separately. Remove transmission to center housing mounting bolts, then carefully pry center housing away from transmission.

To reinstall, reverse the removal procedure. Be sure that shafts are correctly meshed before tightening mounting bolts.

DIFFERENTIAL AND BEVEL GEARS

Models 1100-1110-1200-1210

137. **R&R AND OVERHAUL.** The differential assembly and related parts can be removed from rear axle center housing after removing both final drive (axle) housings as outlined in paragraph 152 and lift arm housing from top of center housing as described in paragraph 180. Bevel pinion (32−Fig. 222) and ring gear (33) are available only as a

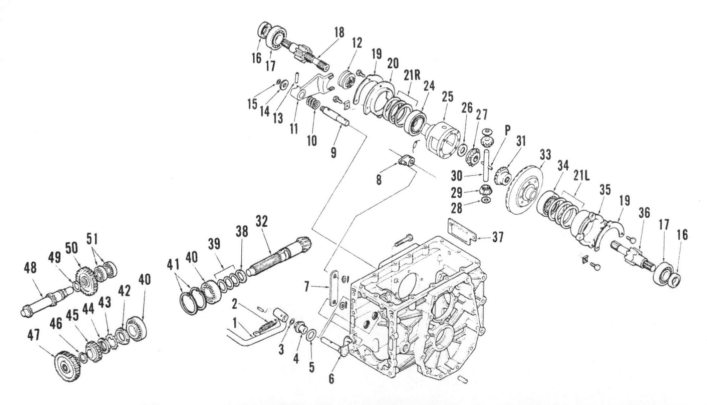

Fig. 222—Exploded view of differential assembly and bevel pinion and ring gear used on 1100, 1110, 1200 and 1210 models. Drive gear (45) is used on models with four wheel drive.

1. Differential lock pedal	9. Shaft	18. Right side pinion
2. Spring	10. Spring	19. Lock plate
3. "O" ring	11. Fork	20. Housing
4. Guide	12. Coupling	21L & 21R. Shims
5. Gasket	13. Cam pin	24. Right carrier bearing
6. Lever	14. Spacer	25. Differential housing
7. Link	15. "O" ring	26. Thrust washer (later
8. Lever	16. Seals	models)
	17. Pinion bearings	27. Right side gear

28. Thrust washers	35. Housing	44. Nut
29. Pinions	36. Left side pinion	45. Front wheel drive
30. Shaft	37. Plate	gear
31. Left side gear	38. Thrust washer	46. Snap ring
32. Bevel pinion	39. Shims	47. Gear
33. Bevel ring gear	40. Bearings	48. Countershaft
34. Left carrier bearing	41. Snap rings	49. Snap ring
	42. Nut	50. Gear
	43. Lockwasher	51. Bearings

matched set. To remove bevel pinion, it is necessary to separate axle center housing from transmission housing as outlined in paragraph 135.

To remove differential assembly after final drive housings and lift housing have been removed, drive pin (13) from differential lock shaft (9) and remove fork (11) and spring (10). Unbolt and remove carrier bearing housings (20 and 35) being careful not to mix or lose shims (21L and 21R). Remove differential assembly out top of center housing.

To remove bevel pinion after center housing is separated from transmission housing, remove drive gear (47–Fig. 222) and snap ring (46). On four wheel drive models, remove gear (45) and key. On all models, straighten tabs of lockwasher (43) and remove nuts (42 and 44). Tap pinion rearward out of front bearing (40) and remove from housing.

To disassemble differential, remove bolts attaching ring gear (33–Fig. 222) to housing (25). Remove side gear (31). Drive the pin (P) from shaft, then remove shaft, pinion gears (29), thrust washers (28), side gear (27) and thrust washer (26) if used.

Inspect all parts for excessive wear or damage. Backlash between pinion gears and side gears should be 0.10-0.15 mm (0.004-0.006 inch) with maximum limit of 0.30 mm (0.012 inch). Thickness of thrust washers (28) when new is 1.0-1.1 mm (0.039-0.043 inch) and wear limit is 0.8 mm (0.031 inch). Clearance between pinion gears (29) and shaft (30) should not exceed 0.5 mm (0.020 inch).

To reassemble, reverse the disassembly procedure. Tighten bevel ring gear mounting bolts to 60-75 N·m (44-55 ft.-lbs.) torque.

Refer to paragraphs 138 through 141 for adjustment of ring gear and pinion.

138. **ADJUSTMENT.** The bevel pinion (32–Fig. 222) and ring gear (33) are a matched set. Be sure that the same assembly number is etched on end of bevel pinion and on circumference of ring gear. The end of bevel pinion is also marked with a "+" or "–" value in millimeters. The +" or "–" value indicates deviation from a nominal value and can be used to help determine proper thickness of shims (39) when renewing bevel pinion. If the value of new pinion is **greater** than value of old pinion, **add** shims (39) equal to this difference. If value of new pinion is **less** than value of old pinion, **remove** shims (39) equal to the difference in values.

139. **BEVEL PINION BEARING PRELOAD.** Install bevel pinion (32–Fig. 222), thrust washer (38) and bearings (40) using correct thickness of shims (39) as determined in paragraph

138. Install nuts (42 and 44) and lockwasher (43). Wrap a cord (Fig. 223) around pinion shaft and attach a spring scale to cord to measure pull necessary to rotate pinion shaft. Tighten inner nut (42–Fig. 222) until a constant pull of 8-10 Kg (18-22 pounds) is required to rotate pinion shaft in bearings. Tighten outer locknut (44), recheck rolling resistance and bend tangs of lockwasher into slots of nuts to lock the adjustment.

140. **DIFFERENTIAL CARRIER BEARINGS ADJUSTMENT.** If differential housing (25–Fig. 222), bevel pinion (32) and ring gear (33), carrier bearings (24 and 34), carrier housings (20 and 35) and axle center housing are all being reused, the original thickness of shims (21L and 21R) should be correct to reinstall in original locations. If any of these parts have been renewed, carrier bearing adjustment and bevel gear backlash should be checked and adjusted as follows:

Install differential assembly in housing selecting shims (21L and 21R) that will provide zero side play of differential carrier bearings. Be sure there is some backlash between ring gear and pinion when measuring differential side play.

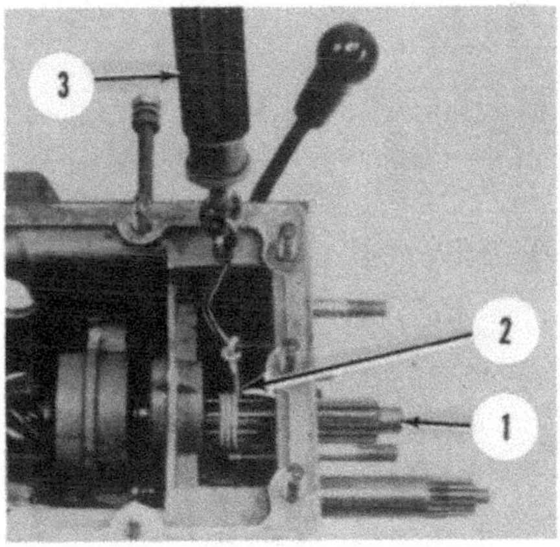

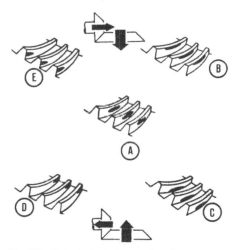

Fig. 223—When adjusting bevel pinion bearing preload, wrap a cord (2) around pinion shaft (1) and use a spring scale (3) to measure pull required to rotate shaft.

Fig. 224—Use a dial indicator (1) to measure backlash between ring gear (2) and bevel pinion.

Do not preload bearings by removing too many shims.

After carrier bearings are correctly adjusted, check backlash between ring gear and pinion using a dial indicator located at right angle to outer edge of ring gear teeth (Fig. 224). Correct backlash is 0.10-0.15 mm (0.004-0.006 inch). If backlash is excessive, move some shims from right side (21R) to left side (21L). Move shims from left side to right side to increase backlash. Do not add or subtract shims from total shim pack thickness as carrier bearing adjustment would be affected. Tighten cap screws attaching carrier housings (20 and 35) to 30-38 N·m (22-28 ft.-lbs.) torque.

141. **BEVEL GEAR MESH POSITION.** As a final assembly check, bevel gear mesh position can be checked as follows: Coat ring gear teeth with Prussian Blue, then turn pinion shaft while applying light resistance to ring gear. Refer to Fig. 225. Desired mesh position is shown at (A).

If mesh position is not correct, check for incorrect assembly of ring gear and pinion. Arrows indicate direction to move gears to correct mesh position. Moving pinion is accomplished by adding

Fig. 225—Refer to text when setting ring gear and pinion gear mesh position.

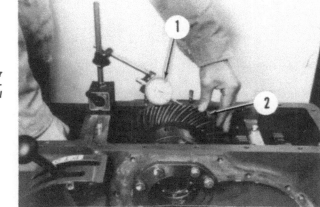

or removing shims (39 – Fig. 222). Moving ring gear is accomplished by transferring shims (21L and 21R) from one side to the other. Be sure mesh position and backlash are correct before finalizing assembly.

Models 1300-1310-1500-1510-1710

142. **R&R AND OVERHAUL.** The differential assembly and related parts can be removed from rear axle center housing after removing both final drive (axle) housings as outlined in paragraph 154 and hydraulic lift housing from top of center housing as described in paragraph 180.

Bevel pinion gear (27 – Fig. 226) and ring gear (17) are available only as a matched set. To remove bevel pinion, it is necessary to separate transmission housing from axle center housing as outlined in paragraph 136.

To remove differential assembly after axle housings and lift housing are removed, proceed as follows: Remove two detent plugs, springs and balls (2 – Fig. 227) from pto shift cover (3), then unbolt and remove shift cover. Drive roll pins from shift fork (10) and boss (7). Slide shift rod (8) out rear of housing while withdrawing shift boss and fork. Unbolt and remove differential carrier housings (9L and 9R – Fig. 226) from center housing while supporting differential assembly. Lift differential assembly from center housing. Be careful not to lose or mix shims (10L and 10R).

To remove bevel pinion (27 – Fig. 227) after center housing is separated from transmission housing, proceed as follows: On models with four wheel drive and/or creeper gear option, slide shift rail (63) out towards the rear while removing shift fork (62) through top opening. Be careful not to lose detent ball and spring (61) as rail is removed. The fork and gears for creeper option are not shown; however, fork is located on rail (63) in front of fork (62), if used, and a gear replaces coupling (31).

On all models, slide coupling (31) and gears (if used) forward on pinion shaft. Straighten tabs of lockwasher (29) and remove nuts (28 and 30). Withdraw bevel pinion (27) with rear bearing (25) towards the rear. Remove front bearing (25) from housing bore.

To disassemble differential, remove cap screws attaching ring gear (17 – Fig. 226) to housing (15). Unbolt and remove housing cover (14), thrust washer (26) and side gear (25). Remove shaft (23) and retainer ball (24), then withdraw pinion gears (22), thrust washers (20) and side gear (19).

Inspect all parts for wear or damage and renew if necessary. Backlash be-

tween pinion gears (22) and side gears (19 and 25) should be 0.10-0.15 mm (0.004-0.006 inch) with a maximum limit of 0.50 mm (0.020 inch). Renew thrust washers (20) if thickness is less than 0.9 mm (0.035 inch). Clearance between pinion gears (22) and shaft (23) should not exceed 0.5 mm (0.020 inch).

To reassemble, reverse the disassembly procedure. Tighten cap screws attaching cover (14) to differential housing to 30-40 N·m (22-29 ft.-lbs.) torque. Tighten cap screws attaching ring rear to differential housing to 60-75 N·m (44-55 ft.-lbs.) torque.

Refer to paragraphs 143 through 146 for adjustment of ring gear and pinion.

143. **ADJUSTMENT.** The bevel pinion (27 – Fig. 226) and ring gear (17) are serviced only as a matched set. Be sure

that the same assembly number is etched on end of pinion and on circumference of ring gear. The end of pinion is also marked with a "+" or "−" value in millimeters. The "+" or "−" value indicates deviation from a nominal value and can be used to help determine proper thickness of shims (24 – Fig. 227) when renewing bevel pinion. If the value of new pinion is **greater** than value of old pinion, **add** shims (24) to original shim pack equal to the difference in values. If value of new pinion is **less** than value of old pinion, **remove** shims from original shim pack equal to the difference.

144. **BEVEL PINION BEARING PRELOAD.** Assemble bevel pinion (27 – Fig. 227), thrust washer (23), shims (24), bearings (25), nuts (28 and 30) and

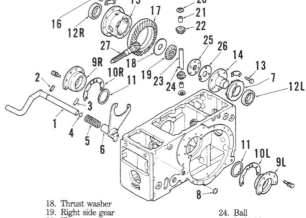

Fig. 226 – Exploded view of differential assembly and related parts typical of 1300, 1310, 1500, 1510 and 1710 models. On 1710 Offset tractors, ring gear (17) is positioned on left side of bevel pinion (27).

1. Differential lock pedal
2. Pin
3. Cam pin
4. "O" ring
5. Spring
6. Shift fork
7. Coupling
8. Plug
9L. Left carrier housing
9R. Right carrier housing
10L & 10R. Shims
11. Snap rings
12L & 12R. Carrier bearings
13. Screws
14. Differential housing cover
15. Differential housing
16. Screws
17. Bevel ring gear
18. Thrust washer
19. Right side gear
20. Thrust washer
21. Bushing
22. Pinion
23. Pinion shaft
24. Ball
25. Left side gear
26. Left side thrust washer
27. Bevel pinion & shaft

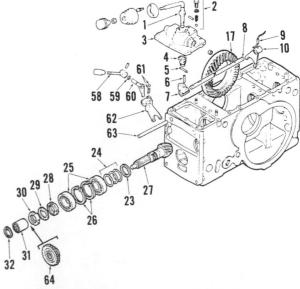

Fig. 227 – Exploded view of bevel drive gears and related parts typical of 1300, 1310, 1500, 1510 and 1710 models. Parts 58 through 64 are used on four wheel drive models. Ring gear (17) is positioned on left side of bevel pinion (27) on 1710 Offset tractors.

1. Pto shift lever
2. Detent assy.
3. Shift cover
4. Spring
5. Pin
6. Roll pin
7. Shift boss
8. Shift rail
9. Pin
10. Pto shift fork
17. Ring gear
23. Thrust washer
24. Shims
25. Bearings
26. Snap rings
27. Bevel pinion
28. Nut
29. Lockwasher
30. Nut
31. Coupling
32. Snap ring
58. Front wheel drive shift lever
59. "O" ring
60. Shift arm
61. Detent spring & ball
62. Front wheel drive shift fork
63. Shift rail
64. Front wheel drive sliding gear

lockwasher (29) in center housing. Wrap a cord (Fig. 223) around bevel pinion shaft and attach a spring scale to measure pull required to rotate pinion shaft. Tighten inner nut (28 – Fig. 227) until a steady pull of 5-6 Kg (11-13 pounds) pull is required to rotate pinion shaft if original bearings (25) were installed. If new bearings were installed, tighten nut until a steady pull of 11-13 Kg (24-29 pounds) is required to rotate shaft in bearings.

After recommended bearing preload is obtained, tighten outer nut (30) and bend tabs of lockwasher (29) into slots of nuts to secure the adjustment.

145. DIFFERENTIAL CARRIER BEARING ADJUSTMENT.

If differential housing (15 – Fig. 226), bevel pinion (27) and ring gear (17), carrier bearings (12L and 12R), carrier housings (9L and 9R) and axle center housing are all being reused, the original thickness of shims (10L and 10R) will probably be correct to reinstall in original locations. If any of these parts have been renewed, carrier bearing adjustment and bevel gear backlash should be checked and adjusted as follows:

Install differential assembly in center housing selecting shims (10L and 10R) that will provide zero side play of differential carrier bearings. Be sure there is some backlash between ring gear and pinion when checking for side play. Do not preload bearings by removing too many shims.

After carrier bearings are correctly adjusted, measure backlash between ring gear and pinion using a dial indicator positioned at right angle to outer edge of ring gear teeth as shown in Fig. 224. Backlash should be 0.10-0.15 mm (0.004-0.006 inch). Move shims (10L and 10R – Fig. 226) from one side to the other to obtain desired backlash. Do not add or remove shims from total shim pack thickness as carrier bearing adjustment would be affected.

146. BEVEL GEAR MESH POSITION.

As a final assembly check, bevel gear tooth contact pattern can be checked as follows: Apply Prussian Blue to ring gear teeth, then turn pinion shaft while applying light resistance to ring gear. Refer to Fig. 225. Desired mesh position is shown at (A).

If mesh position is not correct, check for incorrect assembly of ring gear and pinion. Arrows (Fig. 225) indicate direction to move gears to correct mesh position. Moving bevel pinion is accomplished by adding or removing shims (24 – Fig. 227). Moving ring gear is accomplished by transferring shims (10L and 10R – Fig. 226) from one side to the other. Be sure mesh position and back-

lash are correct before finalizing assembly.

Models 1700-1900-1910-2110

147. R&R AND OVERHAUL. The differential assembly and related parts can be removed from rear axle center housing after removing both final drive (axle) housings as outlined in paragraph 158 and hydraulic lift housing as outlined in paragraph 180.

Bevel pinion gear (27 – Fig. 228 or 229) and ring gear (17) are available only as a matched set. To remove bevel pinion, it is necessary to separate transmission housing from axle center housing as outlined in paragraph 136.

To remove differential assembly after axle housings and lift housing are removed, proceed as follows: Support differential assembly, then remove both carrier housings (9L and 9R – Fig. 228). Lift differential and ring gear from center housing.

To remove bevel pinion after center housing is separated from transmission housing, proceed as follows: Remove cover plate or shift cover from top of center housing if not previously removed. Straighten tabs of lockwasher

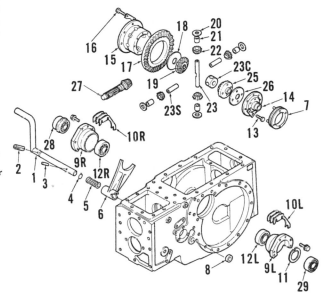

Fig. 228 — Exploded view of differential assembly and related parts used on 1700, 1900, 1910 and 2110 models.

1. Differential lock pedal
2. Roll pin
3. Cam pin
4. "O" ring
5. Spring
6. Fork
7. Coupling
8. Plug
9L. Left carrier housing
9R. Right carrier housing
10L & 10R. Shims
11. Snap ring
12L. Left side carrier bearing
12R. Right side carrier bearing
13. Screws
14. Differential housing cover
15. Differential housing
16. Screws
17. Bevel ring gear
18. Thrust washer
19. Right side gear
20. Thrust washer
21. Bushing
22. Pinion
23. Pinion shaft (long)
23C. Pinion shaft connector
23S. Short pinion shafts
25. Left side gear
26. Left side thrust washer
27. Bevel pinion & shaft
28. Right carrier bearing
29. Left carrier bearing

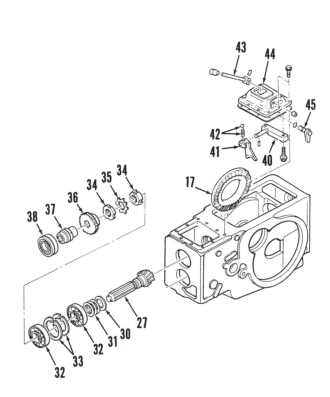

Fig. 229 — Exploded view of bevel drive gears and related parts used on 1910 and 2110 models equipped with four wheel drive. Models 1700 and 1900 are similar. Front wheel drive sliding gear (36) and shifter components (40 through 45) are not used on two wheel drive models.

17. Bevel ring gear
27. Bevel pinion
30. Thrust washer
31. Shims
32. Bearings
33. Snap rings
34. Nuts
35. Lockwasher
36. Front wheel drive sliding gear
37. Coupling
38. Bearing
40. Shift rod
41. Shift fork
42. Detent ball & spring
43. Front wheel drive engage lever
44. Cover
45. Shift arm

(35 – Fig. 229) and remove nuts (34). Withdraw bevel pinion rearward from housing. Remove front bearing (32) from housing bore.

To disassemble differential, unbolt and remove differential housing cover (14 – Fig. 228) and ring gear (17) from differential housing. Remove differential pinion shafts (23 and 23S) thrust washers, pinion gears (22) and side gears (19 and 25).

Inspect all parts for wear or damage and renew if necessary. Clearance between pinion gears and pinion shafts should not exceed 0.5 mm (0.020 inch). Thrust washers should be renewed if thickness is less than 0.9 mm (0.035 inch).

To reassemble, reverse the disassembly procedure. Tighten cap screws attaching ring gear to differential housing to 60-75 N·m (44-55 ft.-lbs.) torque. Tighten housing cover (14) mounting screws to 30-40 N·m (22-28 ft.-lbs.) torque.

Refer to paragraphs 148 through 151 for adjustment of ring gear and pinion.

148. ADJUSTMENT. The bevel pinion (27 – Fig. 228 or 229) and ring gear (17) are serviced only as a matched set. Be sure that the same assembly number is etched on the end of bevel pinion and on circumference of ring gear. The end of pinion is also marked with a "+" or "–" value in millimeters. The "+" or "–" value indicates deviation from a nominal value and can be used to help determine proper thickness of shims (31 – Fig. 229) to be used when renewing bevel pinion. If the value of new pinion is **greater** than value of old pinion, **add** shims (24) to original shim pack equal to the difference in values. If value of new pinion is **less** than value of old pinion, **remove** shims from original shim pack equal to the difference.

149. BEVEL PINION BEARING PRELOAD. Install bevel pinion assembly in center housing. Wrap a cord around bevel pinion shaft and attach a spring scale to measure pull required to rotate pinion in bearings. Tighten inner nut (34 – Fig. 29) until a constant pull of 13-18 Kg (29-38 pounds) for 1910 models or 17-23 Kg (38-50 pounds) for 2110 models is required to rotate pinion shaft.

After recommended bearing preload is obtained, tighten outer nut (34) and bend tabs of lockwasher (35) into slots of nuts to secure the adjustment.

150. DIFFERENTIAL CARRIER BEARING ADJUSTMENT. If differential housing (15 – Fig. 228), bevel pinion (27) and ring gear (17), carrier bearings (12L and 12R), carrier housings (9L and 9R) and axle center housing are all

being reused, the original thickness of shims (10L and 10R) should be reinstalled in original locations. If any of these parts have been renewed, carrier bearing adjustment and bevel gear backlash should be checked and adjusted as follows:

Install differential assembly in center housing selecting shims (10L and 10R) that will provide zero side play of differential carrier bearings. Be sure there is some backlash between ring gear and pinion when checking for side play. Be careful not to preload the bearings.

After carrier bearings are correctly adjusted, measure backlash between ring gear and bevel pinion using a dial indicator positioned at right angle to outer edge of ring gear teeth as shown in Fig. 224. Backlash should be 0.05-0.10 mm (0.002-0.004 inch). Move shims (10L and 10R – Fig. 228) from one side to the other to obtain recommended backlash setting. Do not add or remove shims from total shim pack thickness as carrier bearing adjustment would be affected.

151. BEVEL GEAR MESH POSITION. As a final assembly check, bevel gear tooth contact pattern can be checked as follows: Coat ring gear teeth with Prussian Blue, then turn bevel pinion while applying light resistance to ring gear. Refer to Fig. 225. Desired mesh position is shown at (A).

If mesh position is not correct, check for incorrect assembly of ring gear and pinion. Arrows (Fig. 225) indicate direction to move gears to correct mesh position. Moving bevel pinion is accomplished by adding or removing shims (31 – Fig. 229). Moving ring gear is accomplished by transferring shims (10L and 10R – Fig. 228) from one side to the other. Be sure mesh position and backlash are correct before finalizing assembly.

FINAL DRIVE

Models 1100-1110-1200-1210

152. REMOVE AND REINSTALL. To remove final drive assembly, first drain oil from transmission and rear axle housings. Install wedges between front axle and side frame rails to prevent tipping. Raise rear of tractor and block securely under transmission housing. Remove ROPS frame, fender, rear wheel and hitch lower lift link from side (or sides) which is to be removed. Disconnect brake control rod. Attach a suitable hoist to axle and final drive housing. Remove final drive housing mounting cap screws, then move final

drive assembly away from center housing.

Reinstall final drive housing by reversing the removal procedure.

153. OVERHAUL. To disassemble final drive unit after it is removed, proceed as follows: Remove nut (14 – Fig. 231) from axle shaft. Pull bearing (16), final drive gear (17) and collar (18) from axle, then withdraw axle (23) from bearings (19) and housing. Oil seal (21) can be removed after removing retainer (22).

Unbolt and remove brake cover (4) and brake shoes (8) as an assembly. Remove snap ring (9) and brake drum (10) from pinion shaft (13). Tap pinion shaft and bearing (12) out of axle housing.

Reassemble by reversing disassembly procedure. Tighten axle nut (14) to 79-107 N·m (58-79 ft.-lbs.) torque.

Models 1300-1310-1500-1510-1710

154. REMOVE AND REINSTALL. To remove final drive assembly, first drain oil from transmission and rear axle housings. Install wedges between front axle and side frame rails to prevent tipping. Raise rear of tractor and block securely under transmission housing. Remove ROPS frame, fender, rear wheel and hitch lower lift link from side (or sides) which is to be removed. Disconnect brake control rod. Attach a suitable hoist to axle and final drive housing, remove mounting cap screws and move final drive assembly away from center housing.

Reinstall final drive housing by reversing the removal procedure. Be sure that pto shifter detent spring and ball are in place in rear axle center housing when installing left-hand axle housing.

155. OVERHAUL. To disassemble final drive unit after it is removed, proceed as follows: Remove nut (14 – Fig. 232) from axle shaft. Withdraw final drive gear (17) and spacer (16) from axle shaft. Remove cap screws attaching re-

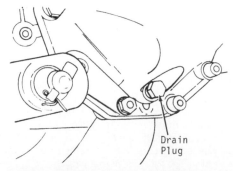

Fig. 230 – Removal of drain plug at bottom of final drive housing will allow fluid to drain from transmission and axle center housing of 1100, 1110, 1200 and 1210 models.

tainer (22) to axle housing, then drive axle shaft outward from housing.

Unbolt and remove brake cover (4) and brake shoes (8) as an assembly from axle housing. Remove snap ring (9) and brake drum (10) from pinion shaft. Remove snap ring (27), then drive pinion shaft (13) out of axle housing.

Reassemble by reversing the removal procedure being careful not to damage oil seals (11 and 21).

Model 1710 Offset

156. **REMOVE AND REINSTALL.** To remove final drive assembly, first

drain oil from transmission housing and final drive housings. Install wedges between front axle and front support rails to prevent tipping. Raise rear of tractor and block securely under transmission housing. Remove ROPS frame, rear wheel and fender. Remove hitch lower lift links and drawbar assembly. On right-hand side, remove cap screws attaching hitch rockshaft support to axle housing. On both sides, disconnect brake control rod from brake cover. Attach a suitable hoist to axle and final drive housing, remove mounting cap screws and move final drive assembly away from center housing.

Reinstall final drive assembly by reversing the removal procedure. Refill final drive housing with Ford 134 or equivalent fluid to oil fill/level check plug opening (Fig. 233).

157. **OVERHAUL.** To disassemble final drive unit after it is removed, remove cap screws attaching final reduction gear housing (47–Fig. 234) to axle housing (33) and separate the housings. Remove cover (32) and seal retainer (31). Remove nut (14) from axle shaft (23). Unbolt and separate housing cover (28) with axle shaft from reduction gear housing (47). Drive the axle shaft out of the cover and remove bearings (19), spacers (16 and 27) and final drive gear (17). Remove final drive pinion (51) and bearings (50 and 52).

Remove retainer (46) from axle housing (33). Remove nut (36) and gear (39) from secondary shaft (42). Drive the secondary shaft outward from the housing.

Remove brake cover (4) with brake shoes (8). Remove snap ring (9) and brake drum (10) from differential output shaft (13). Remove snap ring (53), then drive differential output shaft out of axle housing.

Reassemble by reversing the disassembly procedure.

Models 1700-1900-1910-2110

158. **REMOVE AND REINSTALL.** To remove final drive assembly, first

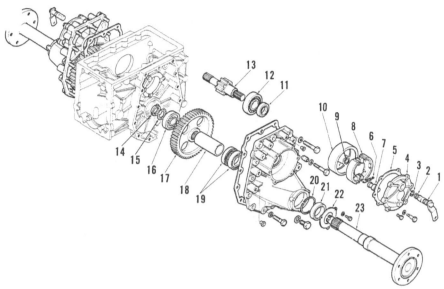

Fig. 231 – Exploded view of final drive and brake used on 1100, 1110, 1200 and 1210 models.

1. Brake cam & lever	7. Anchor pin	13. Drive pinion & shaft	18. Collar
2. "O" rings	8. Brake shoes	14. Nut	19. Bearings
3. Ring	9. Snap ring	15. Lockwasher	20. Snap ring
4. Cover	10. Brake drum	16. Bearing	21. Oil seal
5. Gasket	11. Oil seal	17. Final drive gear	22. Retainer
6. Spring	12. Bearing		23. Axle

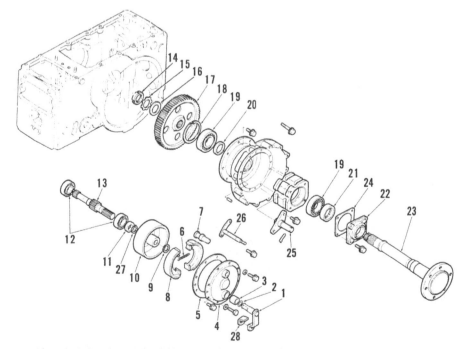

Fig. 232 – Exploded view of final drive assembly used on 1300, 1310, 1500, 1510 and 1710 models.

1. Brake cam & lever	8. Brake shoes	15. Lockwasher	22. Retainer
2. "O" ring	9. Snap ring	16. Washer	23. Axle
3. Bushing	10. Brake drum	17. Final drive gear	24. Gasket
4. Cover	11. Oil seal	18. Snap ring	25. Pin
5. Gasket	12. Bearings	19. Bearings	26. Bracket
6. Spring	13. Drive pinion & shaft	20. Collar	27. Snap ring
7. Anchor pin	14. Nut	21. Oil seal	28. Clamp

Fig. 233 – View of Model 1710 Offset final drive housing oil fill/level check and drain plugs.

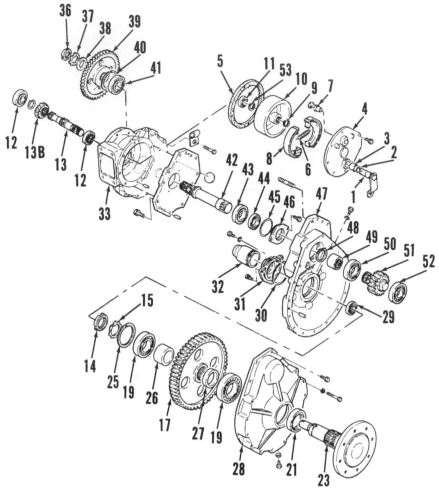

Fig. 234 — Exploded view of final drive assembly used on 1710 Offset tractors.

1. Brake cam & lever	9. Snap ring	17. Final drive gear	29. Oil seal	39. Secondary drive gear	46. Retainer
2. "O" ring	10. Brake drum	19. Bearings	30. "O" ring	40. Snap ring	47. Final drive housing
3. Bushing	11. Oil seal	21. Oil seal	31. Retainer	41. Bearing	48. Snap ring
4. Brake cover	12. Bearings	23. Wheel axle shaft	32. Cover	42. Secondary shaft	49. Coupling
5. Gasket	13. Drive pinion shaft	25. Snap ring	33. Axle housing	43. Bearing	50. Bearing
6. Spring	13B. Pinion gear	26. Spacer	36. Nut	44. Oil seal	51. Final drive pinion
7. Anchor pin	14. Nut	27. Spacer	37. Lockwasher	45. "O" ring	52. Bearing
8. Brake shoes	15. Lockwasher	28. Cover	38. Spacer		53. Snap ring

drain oil from transmission and final drive housings. Install wedges between front axle and frame rails to prevent tipping. Raise rear of tractor and block securely under transmission housing. Remove ROPS frame, rear wheel and fender. Remove hitch lower lift links. Disconnect brake control rod from brake cover. Attach a suitable hoist to axle and final drive housing, remove mounting cap screws and move final drive assembly away from center housing.

Reinstall final drive assembly by reversing the removal procedure. Be sure that pto shift rod detent ball and spring are in place in axle center housing when installing left-hand final drive assembly.

Models 1700-1900-1910 With 152 cm (60 inch) Axle

159. OVERHAUL. Two different axles (23 – Fig. 235) and associated parts have been used. Early 1700 and 1900 models are equipped with a snap ring (14E) which retains final drive gear (17), while later models use a nut (14) to retain the gear. Early models may be updated using later model axle (23), spacers (18), spacer (16), lockwasher (15) and nut (14).

To remove axle from early models, remove snap ring (14E) and final drive gear (17). Unbolt seal retainer (22), then withdraw axle (23) with seal (21), outer bearing (19) and nut (18E) from housing. Remove the nut, then pull bearing from shaft.

To remove axle from later 1700 and 1900 models and 1910 models, remove nut (14) and final drive gear (17) from axle shaft. Unbolt seal retainer (22) and withdraw axle (23) from housing.

Remove brake cover (4) with brake shoes (8). Remove snap ring (9) and brake drum (10) from differential output shaft (13). Remove snap ring, then drive differential output shaft out of axle housing.

Reassemble by reversing the disassembly procedure.

Model 1910 With 183 cm (72 inch) Axle and Model 2110

160. OVERHAUL. To disassemble final drive unit, remove nut (14 – Fig. 235A) and final drive gear (17) from axle shaft. Remove cap screws from seal retainer (22), then drive axle shaft (23) with seal (21) and outer bearings (19) out of housing. Remove snap ring (25), inner bearing (19N) and spacer (26) from housing bore.

Unbolt and remove brake cover (4) and brake shoes (8) as an assembly. Remove snap ring (9) and brake drum (10) from pinion shaft (13). Drive pinion shaft inward and remove from housing.

To reassemble, reverse the disassembly procedure.

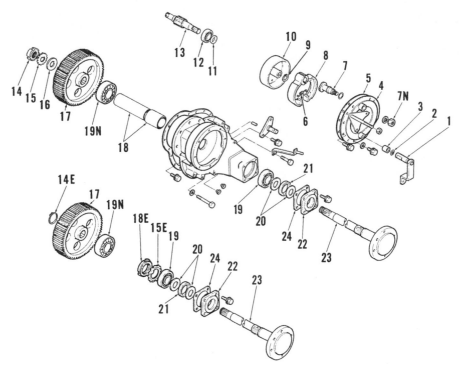

Fig. 235 – Exploded view of final drive assembly used on 1700, 1900 and 1910 models with standard 152 mm (60 inches) tread axle. Parts (14E, 15E and 18E) are used on early 1700 and 1900 models.

1. Brake cam & lever
2. "O" ring
3. Bushing
4. Brake cover
5. Gasket
6. Spring
7. Anchor pin
7N. Nut

8. Brake shoes
9. Snap ring
10. Brake drum
11. Oil seal
12. Bearing
13. Drive pinion
14. Nut

15. Lockwasher
15E. Lockwasher
16. Spacer
17. Final drive gear
18. Spacers
18E. Nut
19. Outer bearing

19N. Inner bearing
20. Spacers
21. Oil seal
22. Retainer
23. Axle shaft
23E. Axle shaft
24. Gasket

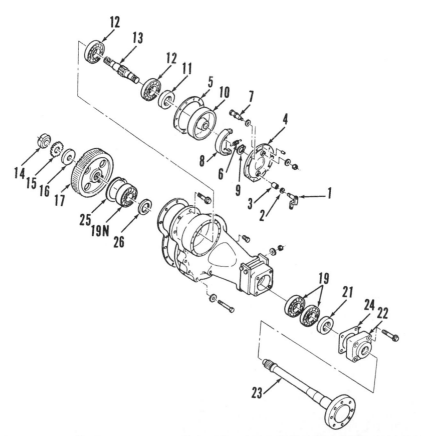

Fig. 235A – Exploded view of final drive assembly typical of 1910 models with 183 cm (72 inch) tread axle and all 2110 models. A single outer axle bearing (19) is used on 1910 models. Refer to Fig. 235 for legend except for snap ring (25) and spacer (26).

BRAKES

ADJUST

All Models

161. Brake pedals should have 20 to 30 mm (¾ to 1-¾ inches) free travel (1–Fig. 236). Adjustment is accomplished by shortening or lengthening brake actuating rods (2). Be sure that braking action is equal on both wheels.

R&R AND OVERHAUL

All Models

162. Refer to Fig. 231, 232, 234, 235 or 235A for an exploded view of brake components. To disassemble, disconnect brake control rod from brake cam lever (1). Unbolt and remove brake cover (4) and brake shoes (8) as an assembly. Snap ring (9) retains brake drum (10) on pinion shaft (13).

Inspect all parts for wear or damage and renew if necessary. If brake compartment is contaminated with oil, seal (11) should be renewed. Refer to the following specification data for all models.

Brake Lining Thickness –
New ....5 mm
(0.197 in.)
Wear Limit ....3.5 mm
(0.138 in.)
Brake Drum ID–
1100, 1110, 1200 and
1210, New................110 mm
(4.331 in.)
Wear Limit.............112 mm
(4.410 in.)
1300, 1310, 1500, 1510,
1700, 1710, 1900 and
1910, New.............150 mm
(5.905 in.)
Wear Limit.............152 mm
(5.985 in.)
2110, New ....200 mm
(7.874 in.)
Wear Limit.............202 mm
(7.953 in.)

POWER TAKE-OFF

Models 1100-1110-1200-1210

163. **STANDARD PTO.** A standard 540 rpm pto (Fig. 237) is available on all tractors. The pto drive is taken from the engine clutch and transmitted through reduction gears in transmission to pto countershaft and output shaft in rear axle center housing. Refer to appropriate Fig. 238 or 239 for exploded view of pto components.

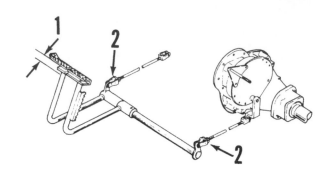

Fig. 236– Brake pedal free travel (1) should be 20 to 30 mm (3/4 to 1-3/16 inches) and is adjusted by lengthening or shortening control rods (2).

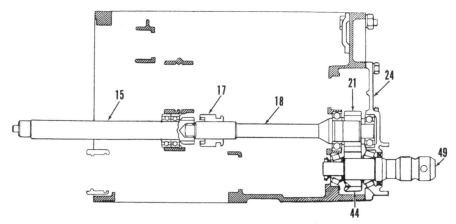

Fig. 237 – Cross section of standard 540 rpm pto available on 1100, 1110, 1200 and 1210 models. Refer to Fig. 238 for legend.

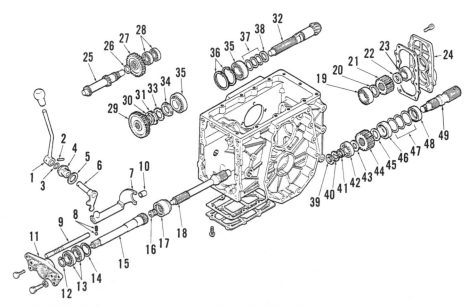

Fig. 238 – Exploded view of standard pto components used on 1100, 1110, 1200 and 1210 two wheel drive models.

1. Pto shift lever	13. Bearings	25. Countershaft	37. Shims
2. Roll pin	14. Snap ring	26. Snap ring	38. Washer
3. "O" ring	15. Pto countershaft	27. Gear	39. Nut
4. Guide	16. Bearing	28. Bearings	40. Lockwasher
5. Gasket	17. Coupling	29. Gear	41. Collar
6. Lever	18. Pto countershaft	30. Snap ring	42. Bearing
7. Shift fork	19. Bearing	31. Nut	43. Collar
8. Detent ball &	20. Collar	32. Bevel pinion shaft	44. Gear
spring	21. Gear	33. Lockwasher	45. Collar
9. Shift rail	22. Collar	34. Nut	46. Bearing
10. Collar	23. Bearing	35. Bearings	47. Shims
11. Housing	24. Cover	36. Snap rings	48. Oil seal
12. Snap ring			49. Pto output shaft

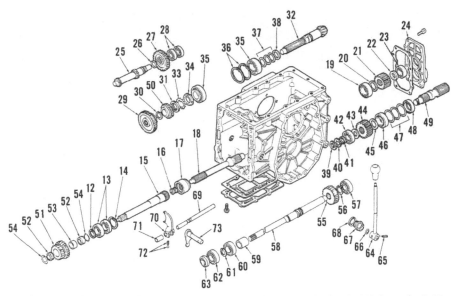

Fig. 239—Exploded view of pto components used on 1100, 1110, 1200 and 1210 four wheel drive models.

12. Snap ring	29. Gear	45. Collar	60. Bearing
13. Bearings	30. Snap ring	46. Bearing	61. Snap ring
14. Snap ring	31. Nut	47. Shims	62. Bearing
15. Pto countershaft	32. Bevel pinion shaft	48. Oil seal	63. Oil seal
16. Bearing	33. Lockwasher	49. Pto output shaft	64. Front wheel engage
17. Coupling	34. Nut	50. Front wheel drive	lever
18. Pto countershaft	35. Bearings	gear	65. Roll pin
19. Bearing	36. Snap rings	51. Gear	66. "O" ring
20. Collar	37. Shims	52. Bearings	67. Guide
21. Gear	38. Washer	53. Spacer	68. Gasket
22. Collar	39. Nut	54. Clips	69. Rail
23. Bearing	40. Lockwasher	55. Gear	70. Shift fork
24. Cover	41. Collar	56. Snap ring	71. Collar
25. Countershaft	42. Bearing	57. Bearing	72. Detent ball and
26. Snap ring	43. Collar	58. Front drive shaft	spring
27. Gear	44. Gear	59. Collar	73. Lever
28. Bearings			

following special instructions. Output shaft bearing preload is adjusted by adding or removing shims (47) between rear bearing (46) and retainer cover (24). Use a spring scale with a cord wrapped around output shaft as shown in Fig. 240 to measure pull required to rotate shaft. Bearing preload is correct when a steady pull of 4-5 Kg (9-11 pounds) is required to turn output shaft.

164. **MID-MOUNT PTO.** An optional mid-mount pto (Fig. 241) is available on 1110 and 1210 models equipped with hydrostatic drive transmission. The mid-mount pto output shaft (6) operates at 1000 rpm, while the pto rear output shaft (8) operates at 540 rpm.

To service mid-mount pto gearbox, first unbolt and remove housing (7 – Fig. 242) from axle center housing. Drive retaining pin (10) out of idler shaft (8). Remove snap ring (15), then withdraw idler shaft rearward from housing and remove idler gear (13), then pull output shaft and bearings from housing.

To service pto front countershafts (21 and 28), the transmission must be separated from rear axle center housing as outlined in paragraph 135. Rear countershaft (31) and sliding gear (30) can be removed from the rear after removing rear pto bearing retainer plate.

When reassembling, be sure that grooved side of thrust washers (11 and 23) face the needle bearings (12 and 24).

Fig. 240—Use a spring scale and cord to measure pull required to rotate pto output shaft (1) when adjusting bearing preload. Refer to text.

The pto output shaft (49), rear countershaft (18) and related gears and bearings can be removed from the rear after removing cover (24) from rear of axle center housing. To remove front countershaft (15) and bearings (13), it is necessary to first separate tractor between transmission and rear axle center housing as outlined in paragraph 135.

To reassemble, reverse the disassembly procedure while noting the

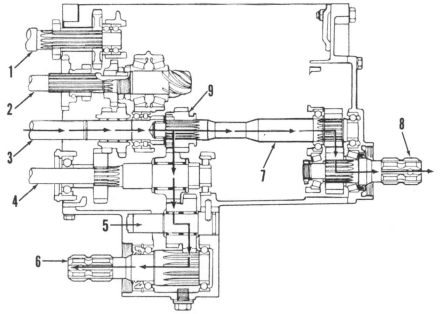

Fig. 241—Cross-sectional view of optional mid-mount pto available on 1110 and 1210 hydrostatic drive models.

1. Transmission drive shaft	3. Pto front countershaft	5. Idler shaft	7. Pto rear countershaft
2. Differential pinion shaft	4. Front wheel drive shaft	6. Mid-mount pto output shaft	8. Rear pto output shaft
			9. Pto sliding coupler

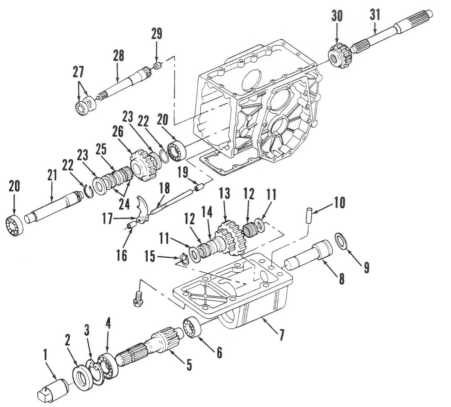

Fig. 242 — Exploded view of mid-mount pto components. On models with front wheel drive, the front wheel drive shaft replaces the countershaft (21).

1. Cover
2. Oil seal
3. Snap ring
4. Bearing
5. Pto output shaft
6. Bearing
7. Housing
8. Idler shaft
9. "O" ring
10. Pin
11. Thrust washers
12. Bearings
13. Idler gear
14. Spacer
15. Snap ring
16. Spacer
17. Shift fork
18. Shift rail
19. Spacer
20. Bearing
21. Countershaft
22. Snap ring
23. Thrust washers
24. Bearings
25. Spacer
26. Countershaft gear
27. Bearing
28. Front countershaft
29. Bearing
30. Sliding gear
31. Rear countershaft

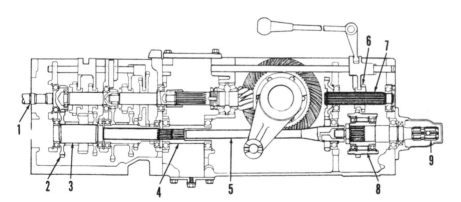

Fig. 243 — Cross-sectional view of transmission drive pto used on 1300, 1310, 1500, 1510 and 1710 models with single plate clutch.

1. Input shaft
2. Counter gear
3. Transmission countershaft
4. Coupling
5. Pto lower countershaft
6. Sliding gear
7. Pto upper countershaft
8. One-way clutch assy.
9. Pto output shaft

Models 1300-1310-1500-1510-1710

165. The pto output shaft oil seal (53 – Fig. 244 or 245) can be renewed after unbolting and removing retainer (55). To remove other components of pto drive, it is necessary to first separate transmission from rear axle center housing as outlined in paragraph 136 and remove hydraulic lift cover as outlined in paragraph 180.

On 1300 and 1500 models, drive roll pins out of shift rail (8 – Fig. 244). Slide shift rail rearward from center housing and remove shifter boss (7) and fork (10).

On 1310, 1510 and 1710 models, drive roll pin (9 – Fig. 245) out of shift rail (8). Thread a 6 mm bolt into retaining pin (6), then withdraw pin from housing bore. Slide shift rail (8) out of center housing and remove shift fork (10).

On all models, remove cover (12 – Fig. 244 or 245) and snap rings (13 and 14) from rear of upper countershaft (20). Remove snap ring (22) and gear (21) from front of countershaft. Tap counter-

shaft rearward and remove front bearing, snap ring (16), spacer (15) and sliding gear (11) as countershaft is withdrawn.

Remove seal retainer (56) from rear of center housing. Withdraw output shaft (49) from housing and remove one-way clutch assembly (nonsynchromesh transmission) or gear (synchromesh transmission) out top opening of housing.

Remove snap ring (36) from groove in rear countershaft (40). Drive countershaft rearward from housing and remove gear (37).

1. Pto shift lever
2. Detent assy. (4 used)
3. Cover
4. Spring
5. Pin
6. Roll pin
7. Shift boss
8. Shift rod
9. Roll pin
10. Pto shift fork
11. Pto sliding gear
12. Cup
13. Snap ring
14. Snap ring
15. Collar
16. Snap ring
17. Bevel ring gear
18. Bearing
18R. Rear bearing
19. Snap ring
20. Pto shaft
21. Gear
22. Snap ring
23. Washer
24. Shims
25. Bearings
26. Snap rings
27. Bevel pinion & shaft
28. Nut
29. Lockwasher
30. Nut
31. Coupling
32. Snap ring
33. Bearing
34. Coupling
35. Snap ring
36. Snap ring
37. Pto gear
38. Bearing
39. Washer
40. Pto drive shaft
41. Bearing

42. Washer
43. Snap ring
44. Bearings
45. Spacers
46. One-way clutch
47. Inner hub
48. Driven gear
49. Pto output shaft
50. Snap ring
51. Bearing
52. Washer
53. Snap ring
54. Oil seal
55. Gasket
56. Retainer
57. Cover
58. Front wheel drive lever
59. "O" ring
60. Lever
61. Detent spring and ball
62. Front wheel drive shift fork
63. Shift rail
64. Front wheel drive sliding gear
65. Spacer
66. Bearings
67. Snap ring
68. Idler gear
69. Shaft
70. Driven gear
71. Bearings
72. Shaft
73. Housing
74. Oil seal
75. "O" rings
76. Holder
77. Coupling
78. Snap ring
79. Front drive shaft
80. Cover

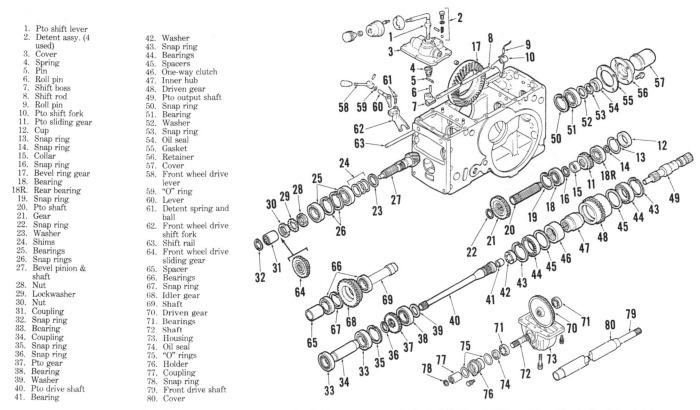

Fig. 244—Exploded view showing pto drive and front wheel drive components typical of 1300 and 1500 models with single plate clutch.

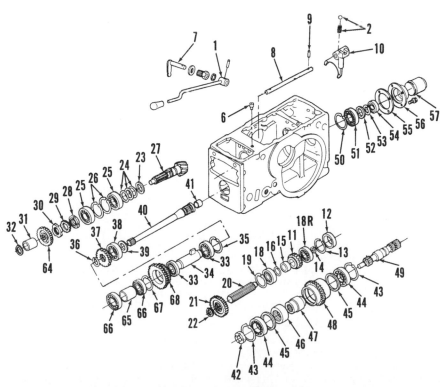

Fig. 245—Exploded view of pto drive components typical of 1310, 1510 and 1710 models. A fixed gear is used in place of one-way clutch assembly (43 through 48) on models with "live" pto. Refer to Fig. 244 for legend.

1. Pto shift lever
2. Detent spring & ball
3. Cover
4. Guide bushing
5. "O" ring
6. Lever
7. Plate
8. Shift rod
9. Roll pin
10. Pto shift fork
11. Gear
12. Cup
13. Spacer
14. Snap ring
15. Spacer
16. Oil slinger (2WD only)
17. Bevel ring gear
18. Bearing
19. Bearing
20. Upper countershaft
21. Gear
22. Snap rings
23. Bearing retainer
24. Shims
25. Bearings
26. Snap rings
27. Bevel pinion & shaft
28. Nut
29. Lockwasher
30. Nut
31. Coupling
32. Bearing
33. Bearing
34. Coupling & snap ring
35. Snap ring
36. Snap ring
37. Gear
38. Bearing
39. Snap rings
40. Pto countershaft
41. Bearing
42. Snap ring
43. Snap ring
43R. Snap ring
44. Bearing
44R. Bearing
45. Spacer
45R. Spacer
46. One-way clutch
47. Clutch hub
48. Sliding gear
49. Pto output shaft
50. Snap ring
51. Bearing
52. Washer
53. Snap ring
54. Oil seal
55. Gasket
56. Retainer
57. Cover
58. Four wheel drive engage lever
59. "O" ring
60. Lever
61. Detent ball & spring
62. Shift fork
63. Shift rod
64. Front wheel drive sliding gear
65. Spacer
66. Bearings
67. Snap ring
68. Idler gear
69. Shaft
70. Driven gear
71. Bearings
72. Shaft
73. Spacer
74. Oil seal
75. "O" rings
76. Bearing holder
77. Coupling
78. Snap ring
79. Front drive shaft
80. Cover
82. Snap ring
83. Bearing
84. Bearing
85. Snap ring

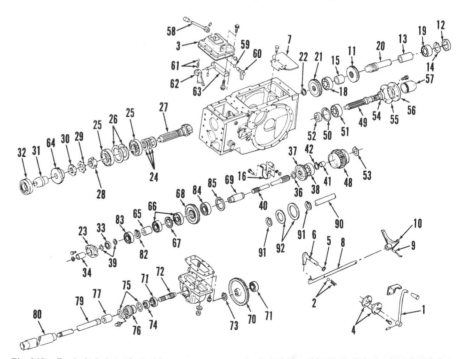

Fig. 248 — Exploded view of pto drive components used on 1700 and 1900 models equipped with four wheel drive. On two wheel drive tractors, components 58 through 85 are not used.

Fig. 249 — Exploded view of pto drive components typical of 1910 and 2110 models equipped with four wheel drive. On two wheel drive tractors, components 58 through 85 are not used. On 1910 models with synchromesh transmission, a one-piece pto countershaft (90) and thrust washers (92) are used in place of countershaft (40), bearing (33), retainer (23) and coupling (34). Refer to Fig. 248 for legend.

To remove coupling (34) and bearings (33), remove snap ring (35) and drive coupling rearward.

Refer to appropriate TRANSMISSION section for service procedures covering front countershaft and input shaft.

To reassemble, reverse the removal procedure.

Models 1700-1900-1910-2110

166. The pto output shaft oil seal (54–Fig. 248 or 249) can be renewed after unbolting and removing retainer (56). To remove other pto drive components located in rear axle center housing, the hydraulic lift cover must be removed as outlined in paragraph 180 and tractor must be split between transmission and center housing as outlined in paragraph 136. Refer to appropriate TRANSMISSION section for service information covering front countershaft and pto input shaft.

To remove upper countershaft (20), remove cup (12) and snap ring (14) from rear of countershaft. Remove bearing retainer plate (7). Remove snap ring (22) and gear (21) from front of countershaft. Drive the shaft rearward and remove front bearing (18), spacers (15 and 13) and gear (11) out top opening.

Remove output shaft seal retainer (56) and oil seal from rear of center housing. Withdraw output shaft (49) from housing and remove one-way clutch assembly (if used) and sliding gear (48) out top of housing.

To remove pto countershaft (90–Fig. 249) on Model 1910 with synchromesh transmission, remove snap rings (91), thrust washers (92) and oil slinger (16),

if so equipped. Disengage snap ring (36) from groove in countershaft. Drive the countershaft rearward from housing and remove snap ring (36) and gear (37).

To remove countershaft (40–Fig. 248 or 249) on all other models, remove bearing retainer (23), snap rings (39) and bearing (33) from front of countershaft. Remove oil slinger (16) if so equipped. Release snap ring (36) from groove in countershaft. Drive countershaft rearward from housing and remove snap ring (36) and gear (37).

To reassemble, reverse the disassembly procedure.

HYDRAULIC LIFT SYSTEM

The hydraulic lift system consists basically of an oil reservoir, hydraulic pump, control valve, lift cylinder and lift links. All models except 2110 are equipped with a single lever, position control hydraulic system as standard equipment. A two lever, position control and draft control system is optional on 1700, 1710, 1900 and 1910 models and standard equipment on 2110 models.

FLUID AND FILTER

All Models

170. The transmission housing and rear axle center housing serve as a common oil reservoir for the hydraulic system. The oil level should be maintained between the mark on the dipstick and lower end of dipstick. On models having dipstick attached to fill plug, do not screw plug into cover when checking oil level.

The manufacturer recommends that oil should be drained every 300 hours of operation and refilled with new Ford 134 or equivalent fluid. The hydraulic system filter should be cleaned or renewed after every 300 hours of operation also.

On 1100, 1110, 1200 and 1210 models, the oil drain plugs are located at bottom of rear axle housings as shown in Fig. 250. The dipstick and filler opening are located on the transmission top cover.

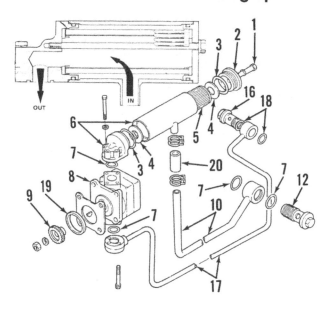

Fig. 251 — The hydraulic system filter (5) is located in housing attached to top of pump (8) on 1100, 1110, 1200 and 1210 tractors (except hydrostatic models).

1. Cover retaining screw
2. Cover
3. "O" ring
4. Seals
5. Filter element
6. Housing
7. "O" rings
8. Pump
9. Drive gear
10. Inlet tube
12. Fitting screw
16. Fitting screw
17. Pressure tube
18. "O" rings
19. Boss
20. Hose

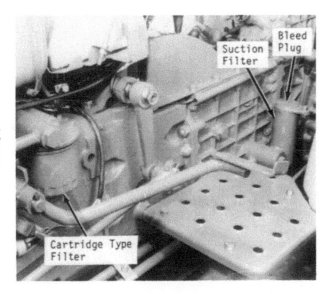

Fig. 252—Two hydraulic system filters are used on 1110 and 1210 models equipped with hydrostatic transmission.

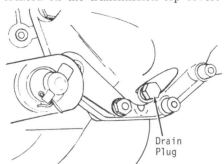

Fig. 250 — View of plug which is removed to drain transmission, rear axle center housing and hydraulic system on 1100, 1110, 1200 and 1210 models.

Hydraulic filter (except hydrostatic transmission models) is located in a housing attached to top of hydraulic pump (Fig. 251). On models with hydrostatic transmission, a spin-on cartridge type filter and a suction filter are located on left side of transmission housing as shown in Fig. 252.

On 1300, 1310, 1500, 1510 and 1710 models, oil drain plugs are located in bottom of transmission housing, each final drive housing and drive housing for front wheel drive if so equipped. Refer to Fig. 253. On 1300 and 1500 models, dipstick is located in transmission top cover and filler opening is located at rear of lift arm housing. On 1310, 1510 and 1710 models, dipstick and filler

Fig. 253 — View of oil drain plugs typical of 1300, 1310, 1500, 1510 and 1710 models.

opening are located in transmission top cover. The hydraulic filter (5–Fig. 254 or 255) is located in housing attached to top of hydraulic pump on 1300, 1310 and

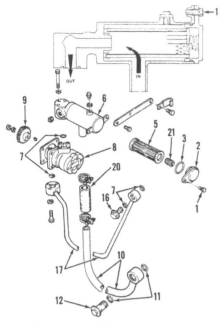

Fig. 254 — Hydraulic filter (5) is located in housing on top of pump on 1300 and 1310 models.

1. Cap screw	7. "O" rings	12. Banjo bolt
2. Cover	8. Pump	16. Banjo bolt
3. "O" ring	9. Drive gear	17. Pressure tube
5. Filter element	10. Inlet tube	20. Hose
6. Housing	11. "O" rings	21. Spring

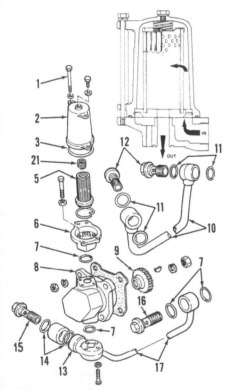

Fig. 255 — Hydraulic system filter (5) is positioned vertically above pump (8) on right front corner of engine on 1500 models. Refer to Fig. 254 for legend except for the following.

13. Lower adapter	15. Banjo bolt
14. "O" rings	16. Banjo bolt

1500 models. On Model 1510, a spin-on type filter is located at suction line flange on left side of rear axle center housing (Fig. 256). On Model 1710, spin-on type filter is mounted on left side rail as shown in Fig. 257.

On 1700 and 1900 models, oil drain plugs are located in bottom of transmission housing, rear axle center housing and drive housing for front wheel drive if so equipped. Refer to Fig. 258. The hydraulic system oil filter is also located in bottom of axle center housing. Note that oil must be drained before removing filter. Dipstick and filter opening are located in transmission top cover.

On 1910 and 2110 models, oil drain plugs are located in bottom of transmission housing, rear axle center housing and drive housing for front wheel drive if so equipped. Refer to Fig. 259. Dipstick and filler opening are located in transmission top cover. The hydraulic system oil filter (1–Fig. 260) is located on top of pump (3) at right front corner of engine. To drain oil from filter prior to removal, remove plug (2) from banjo bolt fitting. With throttle in STOP position, crank engine for about 10 seconds. Remove filter and reinstall plug.

Fig. 256 — A spin-on type hydraulic oil filter (5) is located on left side of rear axle center housing on 1510 models.

Fig. 257 — A spin-on type hydraulic oil filter (5) is located in the pump suction line on left side of engine on 1710 models.

TROUBLE-SHOOTING

All Models

171. The following are symptoms which may occur during operation of the hydraulic lift system and their possible causes. Use this information in conjunction with TESTING and ADJUST-

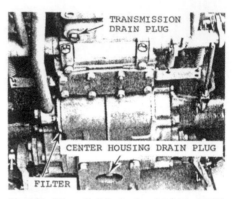

Fig. 258 — View of oil drain plugs typical of 1700 and 1900 models. Hydraulic oil filter is located in the bottom of axle center housing.

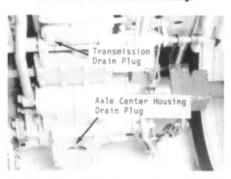

Fig. 259 — View of oil drain plugs typical of 1910 and 2110 models.

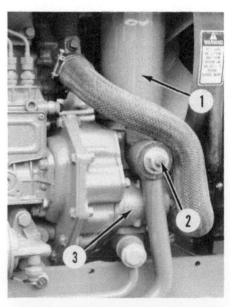

Fig. 260 — Hydraulic system oil filter (1) is located on top of pump (3) on 1910 and 2110 models. Oil can be drained from filter prior to removal by removing plug (2) and cranking engine for about 10 seconds.

MENT information when diagnosing hydraulic lift problems.

1. Hitch will not lift load. Could be caused by:
   a. Linkage out of adjustment or broken.
   b. System relief valve pressure setting too low.
   c. Safety relief valve faulty.
   d. "O" ring failure between control valve and valve cover.
   e. Oil leakage past lift cylinder piston seal.
   f. Oil leakage past unloading valve (single lever system).
   g. Oil leakage past drop poppet valve (single lever system).
   h. Control valve plunger stuck open (two lever system).
   i. Lowering valve spool out of adjustment (two lever system).
   j. Plugged suction filter or low oil level.
   k. Hydraulic pump faulty.
2. Lift arms cycle up and down "hiccup" when control lever is in neutral. Could be caused by:
   a. Poppet valve faulty or misadjusted (single lever system).
   b. Check valve and seat faulty.
   c. Lowering valve spool faulty (two lever system).
3. Hitch will not lower. Could be caused by:
   a. Flow control valve in closed position.
   b. Poppet valve out of adjustment (single lever system).
   c. Lowering valve spool out of adjustment (two lever system).
4. Hitch will not raise to full height. Could be caused by:
   a. Position control feedback rod out of adjustment.
5. System relief valve opens when hitch is in full raise position. Could be caused by:
   a. Position control feedback rod out of adjustment.

TESTING

Models 1100-1200-1300-1500-1700-1710 Offset-1900

172. **RELIEF PRESSURE.** To check system relief valve pressure setting, move position control lever to full lower position and disconnect position control rod (3 – Fig. 261). Remove plug from test port in lift housing cylinder head and install a 0-20000 kPa (0-3000 psi) pressure gage as shown in Fig. 261. Start engine and operate at high rpm. Move control lever (L) to fully raise lift arms and observe pressure reading on gage. Relief valve opening pressure should be 12755 kPa (1850 psi) on 1100,

1200 and 1710 Offset models and 14240-15205 kPa (2065-2205 psi) for all other models.

Relief pressure is adjusted by adding or removing shims (5 – Fig. 263) on

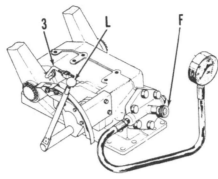

Fig. 261 – On 1100, 1200, 1300, 1500, 1700 and 1900 models, connect pressure gage to cylinder head as shown to check relief valve pressure setting. Refer to text.

F. Flow control valve knob
L. Lift control lever
3. Position control link

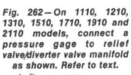

Fig. 262 – On 1110, 1210, 1310, 1510, 1710, 1910 and 2110 models, connect a pressure gage to relief valve/diverter valve manifold as shown. Refer to text.
1. Pressure gage
2. Hose
3. Relief valve
4. Manifold pressure port
5. Diverter valve spool

Fig. 263 – Relief valve assembly at top is typical of type used on 1100, 1200 and 1710 Offset models and is adjusted by adding or removing shims (5). Relief valve at bottom is used on 1300, 1500, 1700 and 1900 models and is adjusted by turning screw (9).
1. Housing
2. Seat
3. Valve
4. Spring
5. Shims
6. "O" ring
7. Plug
8. Housing
9. Adjusting screw
10. Washers
11. Locknut
12. Cap nut

1100, 1200 and 1710 Offset models. One 0.10 mm (0.004 inch) thick shim will change pressure setting approximately 345 kPa (50 psi). On all other models, relief pressure is adjusted by turning adjusting screw (9 – Fig. 263). Be sure pressure setting is correct after locknut (11) is tightened.

Models 1110-1210-1310-1510-1710-1910-2110

173. **RELIEF PRESSURE.** To check system relief valve pressure setting, remove plug from system relief valve/diverter valve manifold and install a 0-20000 kPa (0-3000 psi) pressure gage in manifold pressure port as shown in Fig. 262. Start engine and operate hydraulic system until oil is at normal operating temperature. Set engine speed at high idle, then use a screwdriver to

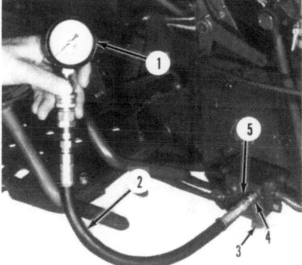

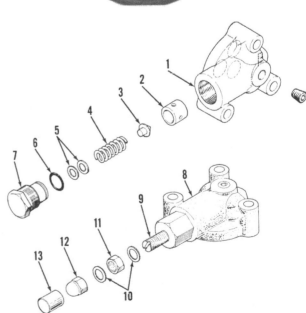

rotate diverter valve spool (5) counter-clockwise and observe pressure gage reading. Relief valve opening pressure should be 12755 kPa (1850 psi) on 1110 and 1210 models. On all other models, relief valve opening pressure should be 14240-15205 kPa (2065-2205 psi).

Relief pressure is adjusted by adding or removing shims (8—Fig. 264) located below relief valve spring (7). One 0.10 mm (0.004 inch) thick shim will change pressure setting approximately 345 kPa (50 psi).

HYDRAULIC PUMP

All Models

174. **REMOVE AND REINSTALL.** To remove hydraulic pump, first clean the pump, fittings and surrounding area to prevent dirt from entering system. On 1500, 1700 and 1900 models, remove lower radiator hose. On 1510 and 1710 models, remove the starting motor. On 1910 and 2110 models, remove plug (2—Fig. 260) from suction tube banjo bolt and crank engine about 10 seconds to drain oil from hydraulic filter. Remove the radiator. On all models, detach inlet and outlet lines from pump, then unbolt pump from engine.

Service parts, other than a seal kit, are not available for repair of hydraulic pumps. If pump components are excessively worn or damaged, renew pump as a complete assembly. Refer to Fig. 265 or Fig. 266 for exploded view of typical hydraulic pumps.

To reinstall pump, reverse the removal procedure.

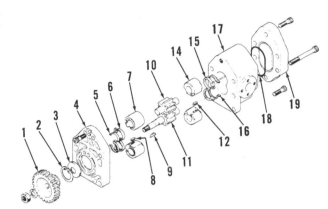

Fig. 265—View of hydraulic pump of type used on some models. Refer also to Fig. 266.
1. Drive gear
2. Snap ring
3. Oil seal
4. Flange
5. Back-up ring
6. Seal
7. Bushing blocks
8. Key
9. Woodruff key
10. Pump gear (driven)
11. Pump gear (drive)
12. Key
14. Bushing blocks
15. Seal
16. Back-up ring
17. Pump body
18. "O" ring
19. Cover

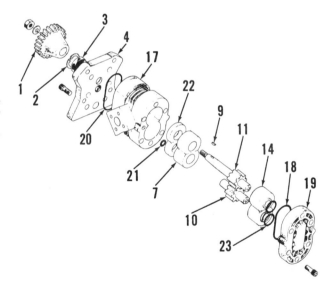

Fig. 266—Exploded view of hydraulic pump of type used on some models. Refer to Fig. 265 for legend except the following.
20. "O" ring
21. "O" ring
22. Spacer
23. "O" ring

ADJUSTMENTS

All Models

175. **POSITION CONTROL ROD.** Length of the position control rod (3—Fig. 267 or Fig. 268) must be adjusted correctly to provide proper operation of hitch. If rod is too short, the con-

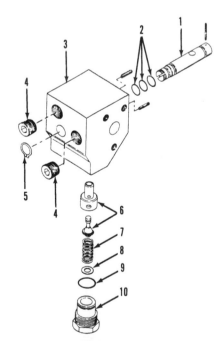

Fig. 264—Exploded view of relief valve/diverter valve assembly used on 1110, 1210, 1310, 1510, 1710, 1910 and 2110 models. Relief pressure setting is adjusted by adding or removing shims (8).
1. Diverter valve spool
2. "O" rings
3. Valve body
4. Plugs
5. Snap ring
6. Relief valve poppet & seat
7. Spring
8. Shim
9. "O" ring
10. Plug

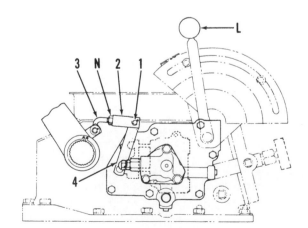

Fig. 267—View of typical single control lever linkage. Refer to text for adjustment procedure.
N. Nut
L. Position control lever
1. Pin
2. Clevis
3. Rod
4. Position control arm

trol will not return to neutral and system relief valve will be actuated when lift arms reach maximum lift position. If rod is too long, full lift height will not be possible.

To adjust, loosen nut (N) and remove pin (1). Move the control lever (L) to the highest lifting position, but not in notch of upper stop. Start engine and permit lift arms to fully raise and relief valve to operate. Move arm (4) clockwise until relief valve just stops opening, then turn clevis (2) until pin (1) can be reinstalled without moving arm (4). Lengthen position control rod one turn (by turning clevis), tighten locknut (N), install pin (1), washer and cotter pin. Check operation to be sure that relief valve does not operate when lift arms reach top of travel.

Models With Draft Control

176. **DRAFT CONTROL ROD.** After position control rod is adjusted correctly, adjust draft control rod as follows: Move position control lever (L–Fig. 268) to highest lifting position, but not in notch at upper end of quadrant. Move draft control lever (D) to upper end of quadrant slot. Loosen locknut (5) on

draft control rod (6), then remove pin (7) attaching control rod to draft control arm (8). Turn flow control valve (F) counterclockwise to full open position.

Start engine and set speed at about 1500 rpm. Move draft control arm (8) rearward until relief valve is actuated, then move arm forward until relief valve stops operating. Adjust length of control rod (6) until pin (7) can be installed without moving control arm, then shorten the rod one full turn. Reconnect rod to arm and move draft control lever to full down position. If arms do not lower or lower too slowly, shorten draft control rod one more turn.

Models With Draft Control

177. **TOP LINK MAIN SPRING.** Length (L—Fig. 269) of top link main spring (1) should be 95-96 mm (3.740-3.780 inches). To adjust spring length, remove cotter pin (3) and turn adjusting nut (2) as necessary.

Models With Draft Control

178. **CONTROL LEVER NEUTRAL ADJUSTMENT.** Start engine and operate at about 1500 rpm. Move posi-

tion control lever (1 – Fig. 270) rearward to approximately the center of quadrant and scribe a line on quadrant at rear edge of control lever. Slowly move lever forward until lift arms just start to lower. Distance (N) lever moves should be 10 mm (7/16 inch).

To adjust control lever neutral travel, the control valve assembly must be removed from lift cover as outlined in paragraph 188. If travel is excessive, turn control valve adjusting screw (2) clockwise to decrease travel distance. If lever travel is too small, turn adjusting screw counterclockwise. Note that insufficient lever neutral travel will cause hitch to "hunt" in neutral position.

All Models

179. **LEVER FRICTION.** The control lever (1 – Fig. 271) should require 2-3 Kg (4.5-6.5 pounds) of pull at lever knob to move the lever. Friction is adjusted by removing cotter pin (2) and turning nut (3) until desired lever friction is obtained.

LIFT ARM HOUSING

All Models

180. **REMOVE AND REINSTALL.** Completely lower lift arms to remove oil from lift cylinder. Remove seat, seat support and fender support rod. Remove rear wiring harness from lift cover clips. On models with draft con-

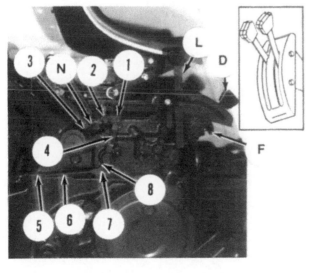

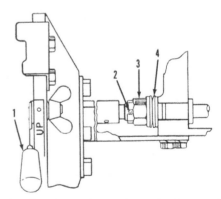

Fig. 268 – View of two control lever linkage typical of models equipped with position and draft control. Refer to text for adjustment procedure.

  D. Draft control lever
  F. Flow control valve
  L. Position control lever
  N. Nut
  1. Pin
  2. Clevis
  3. Position control rod
  4. Position control arm
  5. Nut
  6. Draft control rod
  7. Pin
  8. Draft control arm

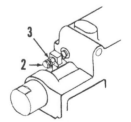

Fig. 271 – View of control lever (1) friction adjusting nut (3), cotter pin (2) and spring washers (4).

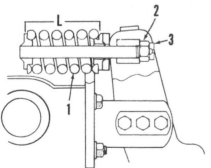

Fig. 269 – On models with draft control, length (L) of top link main spring (1) should be 95-96 mm (3.740-3.780 inches).

Fig. 270 – On models with draft control, position control lever (1) neutral travel (N) should be 10 mm (7/16 inch). Refer to text for adjustment.

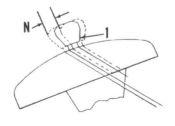

trol, disconnect draft control feedback rod and remove top link bracket assembly. On all models, disconnect high pressure oil lines from lift cover. Disconnect lift links from lift arms. Remove cap screws attaching lift housing to center housing noting different lengths of cap screws and their locations. Use a suitable hoist to remove lift housing from the tractor.

To reinstall, reverse the removal procedure. On 1710 Offset models, be sure to install original shims (5 – Fig. 272) between rockshaft support (6) and support bracket. Be sure that rockshaft rotates freely after support mounting bolts are

tightened. On all models, adjust control linkage as outlined in paragraphs 175 through 179.

181. OVERHAUL LIFT CYLINDER AND ROCKSHAFT. Refer to appropriate Fig. 273, 274, 275, 276 or 277 for exploded view of lift arm housing and related parts.

Prior to removing lift arms (8 and 21) scribe reference marks across lift arms and rockshaft (13) to ensure correct alignment when reassembling. Disconnect position control feedback rod (3). Remove snap rings (7 and 22), then pull lift arms from rockshaft. On models

with snap ring (12) locating rockshaft arm (14) on shaft, disengage snap ring from groove in shaft. On all models, withdraw rockshaft from left-hand side of lift housing.

Remove cylinder head (23) from lift housing, then push piston (16) out front of cylinder.

Inspect all parts for excessive wear or damage and renew if necessary. When renewing bushings (11 and 17), press new bushings in until they are 8 mm (0.315 inch) below flush on 2110 models or 7 mm (0.275 inch) below flush on all other models as shown at (D – Fig. 278). Note that inside diameter of right bushing (11) is smaller than inside diameter of left bushing (17). Renew all "O" rings and seals.

Lubricate all parts with clean hydraulic oil during assembly. Note that one spline on crank arm and one spline on rockshaft has an identification mark (M – Fig. 279) to facilitate correct alignment. Install piston and rod from the front. Be sure to align scribe marks (made during disassembly) on lift arms and rockshaft.

Fig. 272 – View of rear of lift housing on 1710 Offset model equipped with draft control.
1. Top link bracket
2. Draft control bracket
3. Draft control rod
4. Position control link
5. Shim location
6. Rockshaft support

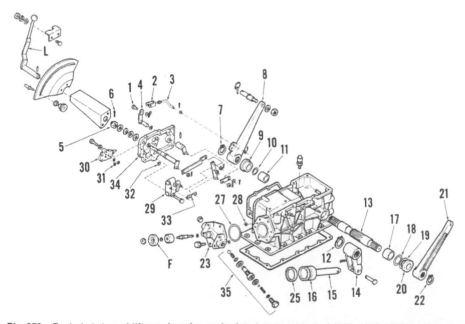

Fig. 273 – Exploded view of lift arm housing and related parts typical of 1100, 1110, 1200 and 1210 models.

F. Flow control valve knob
L. Control lever
1. Pin
2. Clevis
3. Position control link
4. Position control lever
5. Friction adjusting nut
6. Cotter pin
7. Snap ring
8. Right lift arm
9. Collar
10. "O" ring
11. Bushing (30 x 30 mm)
12. Snap ring
13. Shaft
14. Arm
15. Rod
16. Piston
17. Bushing (35 x 30 mm)
18. "O" ring
19. Spring
20. Collar
21. Left lift arm
22. Snap ring
23. Cylinder head
25. Seal ring
27. "O" ring
28. "O" ring
29. Control valve
30. Relief valve
31. "O" rings
32. "O" ring
33. Spring
34. Cover
35. Valves assy.

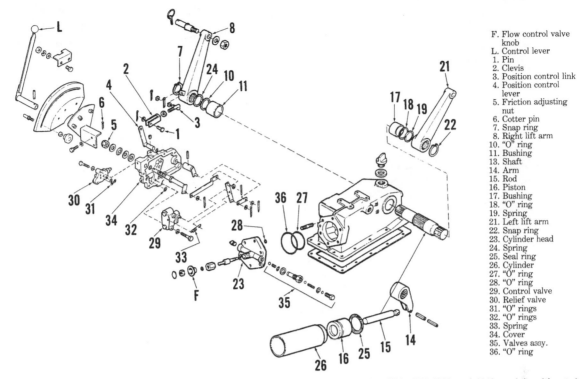

F. Flow control valve
   knob
L. Control lever
1. Pin
2. Clevis
3. Position control link
4. Position control
   lever
5. Friction adjusting
   nut
6. Cotter pin
7. Snap ring
8. Right lift arm
10. "O" ring
11. Bushing
13. Shaft
14. Arm
15. Rod
16. Piston
17. Bushing
18. "O" ring
19. Spring
21. Left lift arm
22. Snap ring
23. Cylinder head
24. Spring
25. Seal ring
26. Cylinder
27. "O" ring
28. "O" ring
29. Control valve
30. Relief valve
31. "O" rings
32. "O" rings
33. Spring
34. Cover
35. Valves assy.
36. "O" ring

Fig. 274—Exploded view of lift arm housing and related parts typical of 1300, 1310, 1500, 1510, 1700, 1710,1900 and 1910 models without draft control.

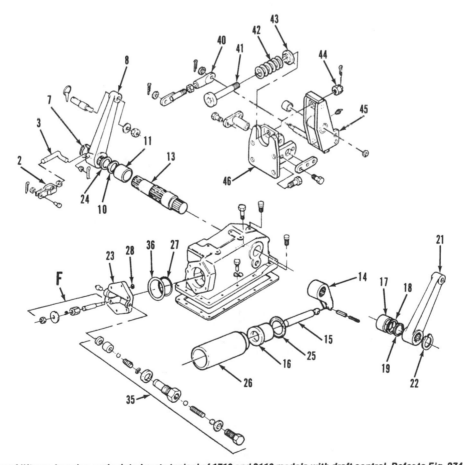

Fig. 275 — Exploded view of lift arm housing and related parts typical of 1710 and 2110 models with draft control. Refer to Fig. 274 for legend except for the following.

40. Draft control feedback rod
41. Pin
42. Main spring
43. Guide
44. Nut
45. Upper link arm
46. Mounting bracket

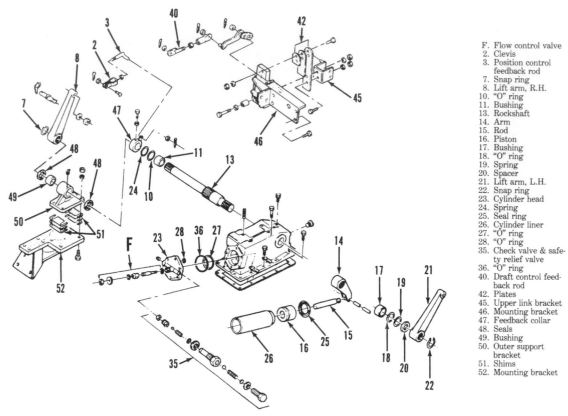

F. Flow control valve
2. Clevis
3. Position control
    feedback rod
7. Snap ring
8. Lift arm, R.H.
10. "O" ring
11. Bushing
13. Rockshaft
14. Arm
15. Rod
16. Piston
17. Bushing
18. "O" ring
19. Spring
20. Spacer
21. Lift arm, L.H.
22. Snap ring
23. Cylinder head
24. Spring
25. Seal ring
26. Cylinder liner
27. "O" ring
28. "O" ring
35. Check valve & safe-
    ty relief valve
36. "O" ring
40. Draft control feed-
    back rod
42. Plates
45. Upper link bracket
46. Mounting bracket
47. Feedback collar
48. Seals
49. Bushing
50. Outer support
    bracket
51. Shims
52. Mounting bracket

Fig. 276 — Exploded view of lift arm housing and related parts used on 1710 Offset model with draft control.

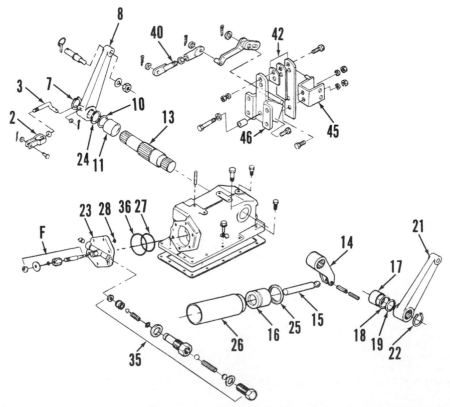

Fig. 277 — Exploded view of lift arm housing and related parts used on 1910 models with draft control. Refer to Fig. 276 for legend.

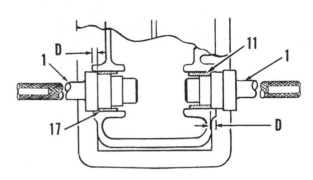

Fig. 278—Use a suitable bushing driver (1) to install rockshaft bushings (11 and 17) to specified depth (D) in lift housing. Refer to text.

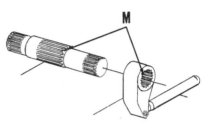

Fig. 279—One spline on rockshaft and one spline on arm has an identificatin mark (M) to facilitate correct alignment.

The safety relief valve (3) protects the lift cylinder against shock loads when the main control valve is in "neutral" position. When lift cylinder pressure exceeds approximately 25000 kPa (3600 psi), the safety valve opens and oil flows from cylinder directly to sump.

To disassemble, unscrew flow control valve knob (1 – Fig. 281), nut (2) and needle (3) from cylinder head (4). Remove relief valve guide bolt (16), shims (14), spring (13) and ball making sure that shims are retained for reassembly. Remove relief valve seat (11), spring (8), check valve ball (7) and seat (6).

Reassemble using new "O" rings and seals.

CONTROL VALVE

Models Without Draft Control

185. **OPERATION.** The control valve assembly contains the control valve

FLOW CONTROL VALVE, CHECK VALVE AND SAFETY RELIEF VALVE

All Models

182. During the "raise" cycle, pressure oil from the pump flows past the flow control valve needle and seat (1 – Fig. 280) and also opens the check valve (2) to enter lift cylinder and raise the lift linkage. During the "lower" cycle, the check valve is closed forcing all return oil from cylinder to flow past the flow control valve needle and seat. Thus, lift linkage lowering speed is controlled by adjusting flow control valve needle to seat clearance. When the flow control valve is fully closed, oil is trapped in lift cylinder and lift arms will remain at their set height.

Fig. 281—Exploded view of flow control valve, safety relief valve and check valve components. Renewable check valve seat (6) and gasket (5) are not used in early models.

1. Flow control knob
2. Nut
3. Flow control valve needle
4. Cylinder head
5. Gasket
6. Check valve seat
7. Check valve ball
8. Spring
9. "O" ring
10. Seal
11. Spring guide & relief valve seat
12. Relief valve ball
13. Spring
14. Shim
15. Seal
16. Spring guide bolt

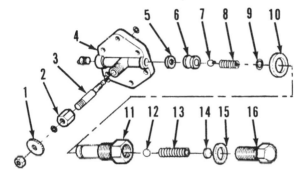

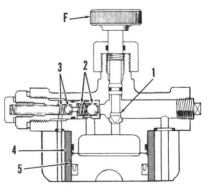

Fig. 280—Cross section of cylinder head showing flow control valve, check valve and safety relief valve typical of all models.

F. Flow control knob
1. Flow control valve needle
2. Check ball & spring
3. Safety relief ball & spring
4. Lift cylinder
5. Piston

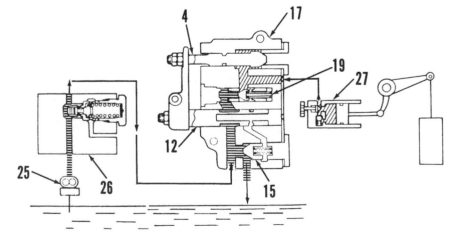

Fig. 282—Cross section of single lever position control valve showing oil flow in "neutral" position.

4. Drop poppet valve
12. Control valve spool
15. Unloading valve
17. Control valve assy.
19. Check valve
25. Pump
26. Relief valve
27. Lift cylinder

spool (12–Fig. 286 or 287), load check valve (19), unloading valve (15) and drop poppet valve (4).

In "neutral" position (Fig. 282), the control valve spool (12) is centered in valve body. Pump oil flow is directed to front of unloading valve (15), which opens and returns pump flow to reservoir. Oil in the lift cylinder is trapped by the check valve (19) and poppet valve (4),

holding the lift arms in a fixed position.

In "lifting" position, the control valve spool (12) is moved inward directing pump oil flow to front and rear of unloading valve (15). Spring pressure then seats the unloading valve, closing the return to reservoir passage. When pump oil pressure (directed to front of check valve) overcomes oil pressure in lift cylinder (directed against rear of check

valve), the check valve (19) opens allowing oil to flow to the lift cylinder. Feedback linkage, connected to the lift arms, returns the control valve spool to "neutral" position when selected lift linkage height is obtained.

In "lowering" position, the control valve spool (12) and poppet valve (4) are both moved outward. Pump oil flow unseats the unloading valve (15) and returns to reservoir. Oil in lift cylinder flows past the unseated poppet valve allowing the lift arms to lower.

186. R&R AND OVERHAUL. The control valve assembly (7–Fig. 285) is mounted on the valve cover (4) inside the lift housing. To remove valve, first fully lower lift linkage to release trapped oil. Disconnect position control feedback rod from feedback lever (3). Disconnect high pressure oil line from valve cover. Unbolt and remove valve cover with control valve from lift housing.

Refer to appropriate Fig. 286 or Fig. 287 for an exploded view of control valve assembly. To disassemble, disconnect linkage return spring (8). Remove nuts (10 and 22) and plate (9) from valve spools and withdraw control valve spool (12). Remove check valve seat (21) and check valve (19). Remove poppet valve seat (1) and poppet valve spool (4). Remove plug (11) and withdraw unloading valve (15).

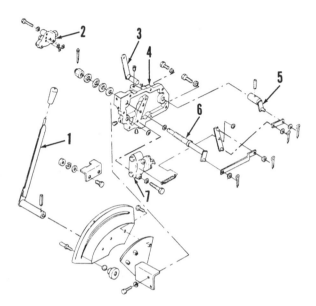

Fig. 285 – Exploded view of control valve linkage typical of single lever position control models.
1. Position control lever
2. Relief valve
3. Feedback lever
4. Cover
5. Feedback arm
6. Position control arm
7. Control valve assy.

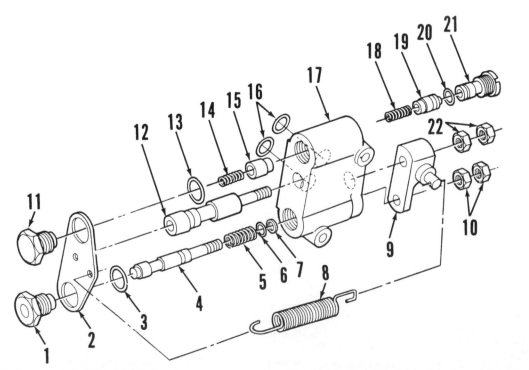

1. Poppet valve seat
2. Plate
3. "O" ring
4. Poppet valve spool
5. Spring
6. "O" ring
7. Back-up ring
8. Spring
9. Valve spool plate
10. Locknuts
11. Unloading valve plug
12. Control valve spool
13. "O" ring
14. Spring
15. Unloading valve
16. "O" rings
17. Valve body
18. Spring
19. Check valve
20. "O" ring
21. Check valve seat
22. Locknuts

Fig. 286 – Exploded view of single lever control valve assembly used on 1100, 1110, 1200, 1210, 1300, 1500, 1700, 1900 models and early production 1310, 1510, 1710 and 1910 models.

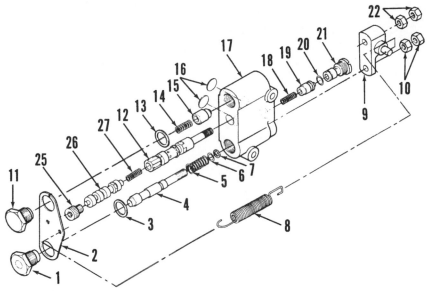

Fig. 287—Exploded view of single lever control valve assembly used on late production 1310, 1510, 1710 and 1910 models. Control valve is similar to early production valve shown in Fig. 286 except for redesigned control valve spool (12) which incorporates an internal spool (26) and spring (27) and a plug (25).

engine rpm. The new style valve can be substituted for old style valve (Fig. 286) if desired.

Inspect surfaces of control valve spool, poppet valve spool, unloading valve, check valve and valve body bores and seats for scratches or excessive wear and renew if necessary. Control valve spool (12) and valve body should be renewed as a set. Be sure to renew all "O" rings.

Lubricate all parts with clean hydraulic oil and reassemble in reverse of disassembly procedure. Outer edge of check valve seat (21) should be staked to valve body after it is installed to lock it in place.

After control valve is reassembled, locate neutral position of control spool (12 – Fig. 288) as follows: Adjust control valve spool locknuts (22) to obtain clearance (A) of 10.15-10.35 mm (0.400-0.407 inch) between valve spool plate (9) and face of valve body. Then, adjust poppet valve nuts (10) to provide clearance (B)

Note that a redesigned control valve spool (12 – Fig. 287), which incorporates an internal spool (26) and spring (27), is used on late production models 1310 (after S.N. UE 03063), 1510 (after S.N. UH 03438), 1710 (after S.N. UL10036) and 1910 (after S.N. UP 05890). The new valve spool assembly provides smoother hydraulic operation at high

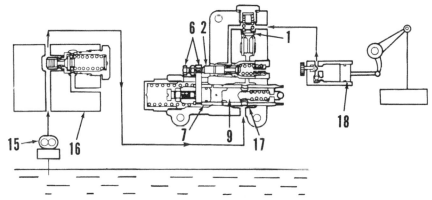

Fig. 290 – Cross section of control valve assembly used on models with draft control showing oil flow in "neutral" position.

1. Check valve	7. Control valve spool	15. Pump	17. Oil passage ports
2. Lowering valve spool	9. Plunger	16. Relief valve	18. Lift cylinder
6. Plate & adjusting screw			

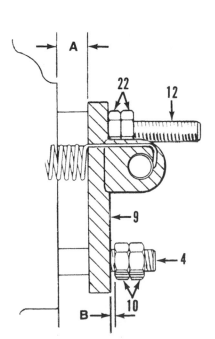

Fig. 288—Control valve spool (12) and poppet valve spool (4) must be adjusted to provide clearance (A) of 10.15-10.35 mm (0.400-0.407 inch) and clearance (B) of 0.30 mm (0.012 inch). Refer to text.

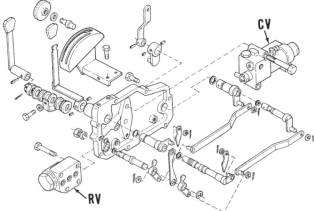

Fig. 291—Exploded view of control valve linkage typical of models with draft control. Relief valve (RV) is used on 1700, 1710 Offset and 1900 models.

of 0.30 mm (0.012 inch) between inner nut and valve plate.

Reinstall control valve and adjust linkage as outlined in paragraph 175.

Models Equipped With Draft Control

187. **OPERATION.** The control valve assembly contains the control valve spool assembly (7–Fig. 292), lowering valve (2) and load check valve (1).

In "neutral" position (Fig. 290), the control valve spool (7) is centered in valve body. Pump oil flow moves the control valve spool plunger (9), which aligns passages (17) in control valve spool and plunger and allows pump oil flow to return to reservoir. Oil is trapped in the lift cylinder by the check valve (1) and lowering valve spool (2), holding the lift arms in a fixed position.

In "lifting" position, the control valve spool (7) is pushed inward. The oil passages in control valve spool and plunger are no longer in alignment which closes the return to reservoir passages. Pump oil flow goes around control valve spool (7) and lowering valve spool (2) to the load check valve (1). When pump pressure overcomes oil pressure in lift cylinder, the check valve opens allowing oil to flow to the lift cylinder. Feedback linkage returns the control valve spool to "neutral" position when selected lift linkage height or draft load setting is obtained.

In "lowering" position, the control valve spool (7) is pulled outward. The passages in control valve spool and plunger (9) are aligned and pump oil flow returns to reservoir as in "neutral." The pin (6) and adjusting screw, which are attached to control valve spool, contact the lowering valve spool (2) and push it off its seat. Oil in lift cylinder flows past the check valve (1) and the lowering valve spool and seat and returns to reservoir.

188. **R&R AND OVERHAUL.** The control valve assembly (CV–Fig. 291) is mounted on the valve cover inside the lift housing. To remove valve, first fully lower lift linkage to discharge trapped oil from the valve. Disconnect control linkage and hydraulic lines from the valve as necessary. Unbolt and remove valve cover with control valve from lift housing.

Refer to Fig. 292 for an exploded view of control valve assembly. To disassemble, unbolt and remove valve from cover. Remove cap (3) and spring from valve body. Remove screw (13) and plug (14) from end of valve spool plug (4), then withdraw pin (6). Do not disturb ad-

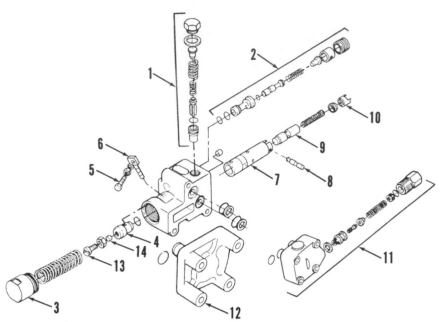

Fig. 292 — Exploded view of control valve assembly used on models with draft control. Relief valve (11) is used on 1700, 1710 Offset and 1900 models; valve plate (12) is used on 1710, 1910 and 2110 models.

1. Check valve assy.
2. Lowering valve assy.
3. Cap
4. Plug
5. Adjusting screw
6. Pin
7. Control valve spool
8. Retaining pin
9. Plunger
10. Spring retainer
11. Relief valve assy.
12. Valve plate
13. Screw
14. Plug

justing screw (5), unless necessary, as control valve adjustment setting will be affected. Remove control valve spool (7) assembly from valve body. Push inward on spring retainer (10) and remove retainer pin (8). Remove plunger (9) from valve spool. Remove check valve assembly (1) and lowering valve assembly.

Inspect valve spools, bores and seats for scratches or excessive wear and

Fig. 295 — Exploded view of remote control valve assembly available on some models equipped with position control hydraulic system.

1. Cap
2. Retaining screw
3. Spring seat
4. Centering spring
5. Plate
6. Wiper
7. "O" ring
8. Valve spool
9. Bracket
10. Steel ball
11. Poppet

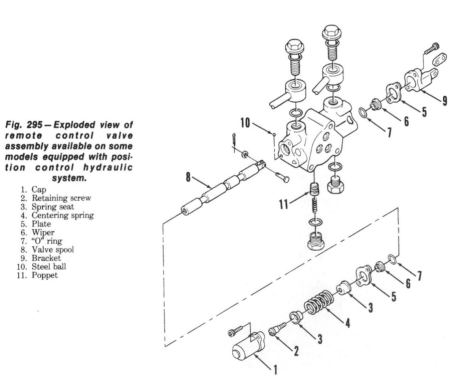

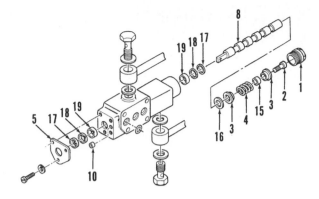

Fig. 296—Exploded view of remote control valve assembly available on some models equipped with draft control hydraulic system. Refer to Fig. 295 for legend except for the following.

10. Plug
15. Spacer
16. Washer
17. Washer
18. Ring
19. Seal

renew if necessary. Control valve spool (7) assembly and valve body must be renewed as a set. Renew all "O" rings.

Lubricate all parts with clean hydraulic oil during assembly. Tighten lowering valve retaining plug to 24 N·m (18 ft.-lbs.) torque. Tighten check valve retaining plug to 58 N·m (43 ft.-lbs.) torque.

Reinstall control valve and adjust linkage as outlined in paragraphs 175 through 178.

REMOTE CONTROL VALVE

All Models So Equipped

190. Refer to appropriate Fig. 295 or Fig. 296 for an exploded view of remote control valve assembly. To disassemble, unbolt and remove bracket (9) and plate (5). Remove end cap (1–Fig. 295) or plug (1–Fig. 296), then withdraw valve spool (8) with centering spring (4) from valve body.

Inspect valve spool and bore for wear, scratches or other damage. Valve spool and body must be renewed as an assembly. Renew all "O" rings and seals.

Lubricate all parts with clean hydraulic oil during assembly.

MID-MOUNT LIFT

A mid-mount lift is used in addition to three-point rear lift system on 1710 Offset tractor. The mid-mount lift is actuated by a double acting remote cylinder and controlled by a single spool remote control valve. Refer to paragraph 190 for repair procedures covering remote control valve.

Model 1710 Offset

191. **REMOTE CYLINDER.** To disassemble, remove nut from cylinder stop rod (28–Fig. 297). Extend the cylinder and remove stop rod and bracket

(29). Remove nuts from cylinder rod bolts and separate cylinder ends (1 and 24) from cylinder barrel (9). Remove piston nut (4) and piston (5) from cylinder rod, then withdraw rod from cylinder end. Remove plate (16), guide plug (17) and poppet (22) from cylinder end.

Inspect all parts for excessive wear, scoring or other damage and renew if necessary. Renew all "O" rings and seals.

To reassemble, reverse the disassembly procedure. Lubricate all parts with clean hydraulic oil during assembly.

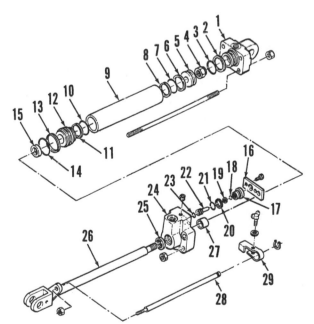

Fig. 297—Exploded view of hydraulic cylinder used on Model 1710 Offset mid-mount lift system.

1. Cylinder end	9. Cylinder barrel	16. Plate	23. "O" ring
2. Back-up ring	10. "O" ring	17. Guide plug	24. Cylinder end
3. "O" ring	11. Back-up ring	18. "O" ring	25. Seal
4. Nut	12. Spacer	19. Washer	26. Piston rod
5. Piston	13. Back-up ring	20. Back-up ring	27. Bushing
6. Back-up ring	14. "O" ring	21. "O" ring	28. Stop rod
7. "O" ring	15. Seal	22. Poppet	29. Stop bracket
8. Back-up ring			

WIRING DIAGRAMS

195. Refer to appropriate Fig. 300 through Fig. 311 for tractor wiring diagram. Refer to paragraphs 90 through 98 for service procedures covering electrical system components.

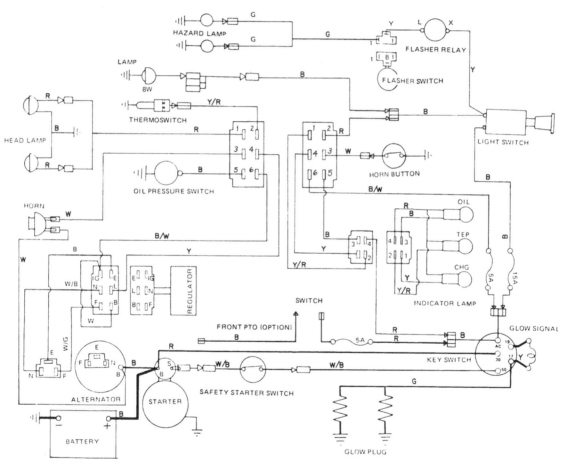

Fig. 300 — Wiring diagram typical of 1100 and 1200 models. Wire color code is as follows:

B. Black	R. Red	B/W. Black/white stripe
G. Green	W. White	W/B. White/black stripe
L. Blue	Y. Yellow	Y/R. Yellow/red stripe

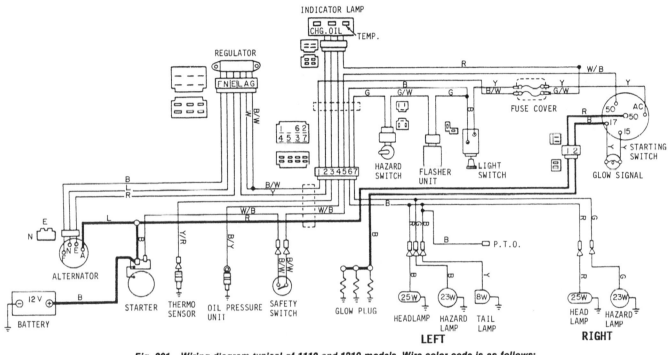

Fig. 301—Wiring diagram typical of 1110 and 1210 models. Wire color code is as follows:

B. Black
BL. Blue
G. Green
R. Red

W. White
Y. Yellow
B/W. Black/white stripe

B/Y. Black/yellow stripe
G/W. Green/white stripe
W/B. White/black stripe

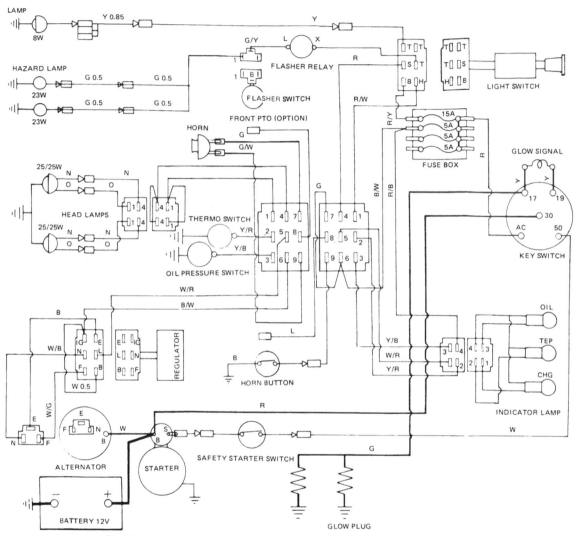

Fig. 302 — Wiring diagram typical of 1300 models. Wire color code is as follows:

B. Black	R. Red	G/Y. Green/yellow stripe	W/G. White/green stripe
G. Green	W. White	R/B. Red/black stripe	W/R. White/red stripe
L. Blue	Y. Yellow	R/W. Red/white stripe	Y/B. Yellow/black stripe
N. Brown	B/W. Black/white stripe	R/Y. Red/yellow stripe	Y/R. Yellow/red stripe
O. Orange	G/R. Green/red stripe	W/B. White/black stripe	

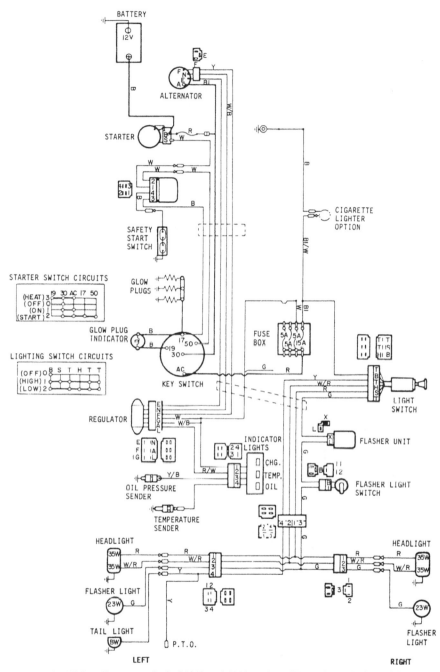

Fig. 303—Wiring diagram typical of 1310 and 1510 models. Wire color code is as follows:

B. Black	W. White	R/W. Red/white stripe	W/R. White/red stripe
BL. Blue	Y. Yellow	W/B. White/black	Y/B. Yellow/black
G. Green	BL/W. Blue/white stripe	stripe	stripe
R. Red			

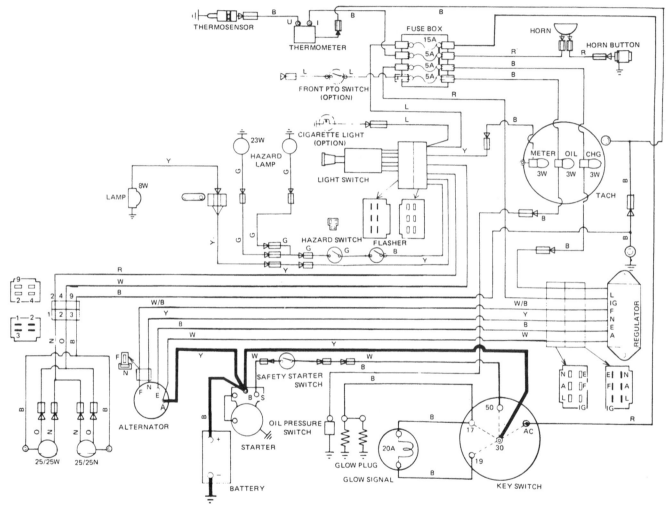

Fig. 304 — Wiring diagram typical of 1500 models. Wire color code is as follows:

B. Black	N. Brown	R. Red	Y. Yellow
G. Green	O. Orange	W. White	W/B. White/black stripe
L. Blue			

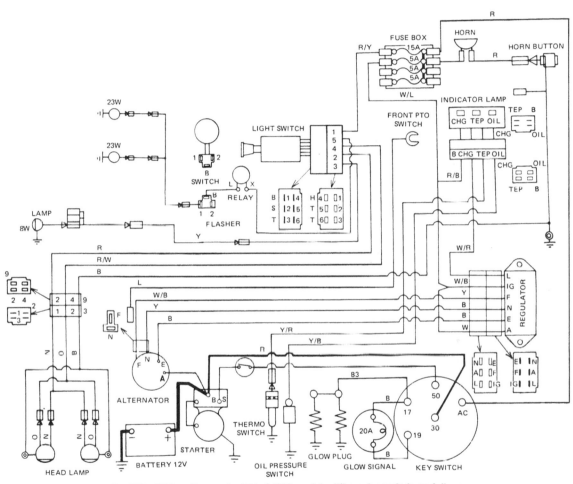

Fig. 305 — Wiring diagram typical of 1700 models. Wire color code is as follows:

B. Black	W. White	R/Y. Red/yellow stripe	W/R. White/red stripe
L. Blue	Y. Yellow	W/B. White/black stripe	Y/B. Yellow/black stripe
N. Brown	R/B. Red/black stripe	W/L. White/blue stripe	Y/R. Yellow/red stripe
O. Orange	R/W. Red/white stripe		
R. Red			

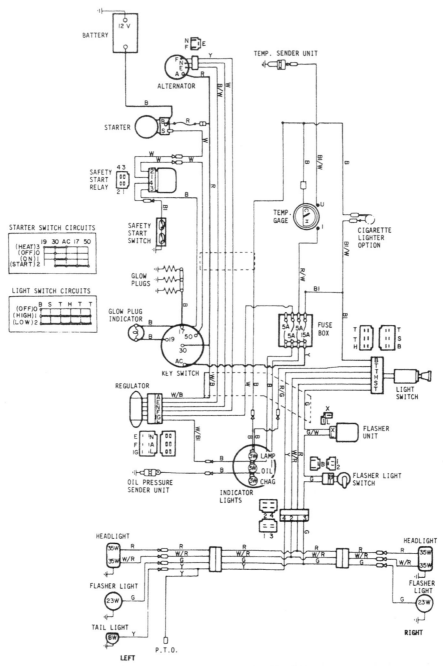

Fig. 306 — Wiring diagram typical of early production (prior to July 1985) 1710 models. Wire color code is as follows:

B. Black			
BL. Blue		G/W. Green/white	W/B. White/black
G. Green	Y. Yellow	stripe	stripe
R. Red	B/W. Black/white	R/G. Red/green stripe	W/BL. White/blue stripe
W. White	stripe	R/W. Red/white stripe	W/R. White/red stripe
	BL/W. Blue/white stripe		

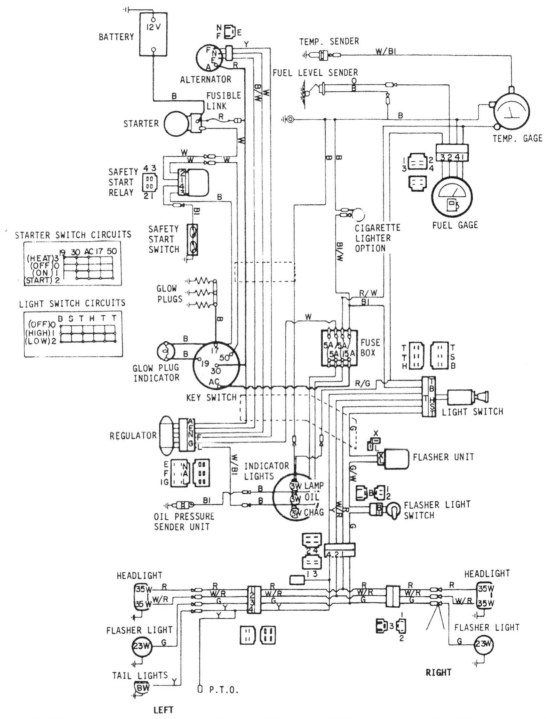

Fig. 307 — Wiring diagram typical of late production (July 1985 and after) 1710 models. Refer to Fig. 306 for wire color code.

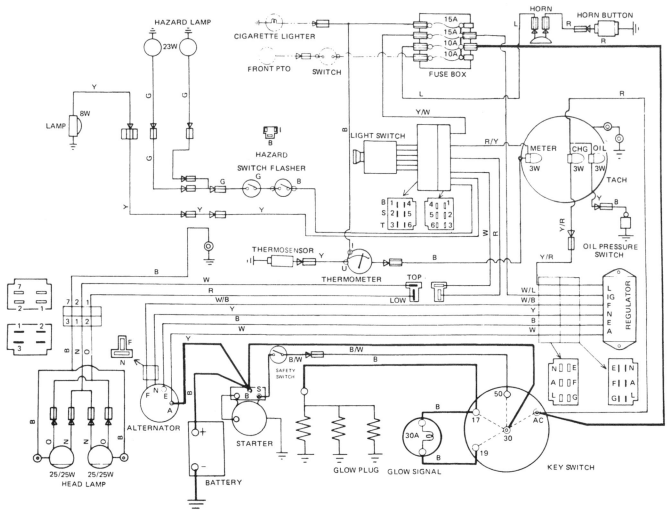

Fig. 308 — Wiring diagram typical of 1900 models. Wire color code is as follows:

B. Black	O. Orange	B/W. Black/white stripe	W/L. White/blue stripe
G. Green	R. Red		R/Y. Red/yellow stripe
L. Blue	W. White	W/B. White/black stripe	Y/R. Yellow/red stripe
N. Brown	Y. Yellow		Y/W. Yellow/white stripe

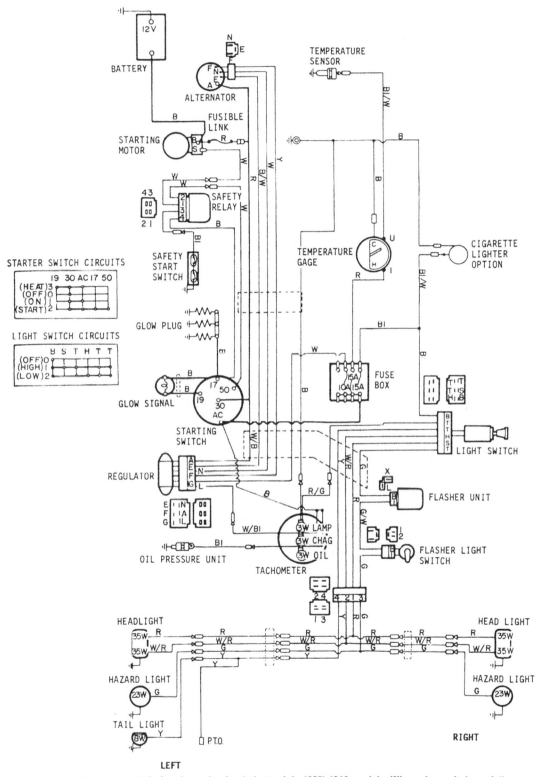

Fig. 309 — Wiring diagram typical of early production (prior to July 1985) 1910 models. Wire color code is as follows:

B. Black	W. White	BL/W. Blue/white stripe	W/B. White/black stripe
BL. Blue	Y. Yellow	G/W. Green/white stripe	W/BL. White/blue stripe
G. Green	B.W. Black/white stripe	R/G. Red/green stripe	W/R. White/red stripe
R. Red			

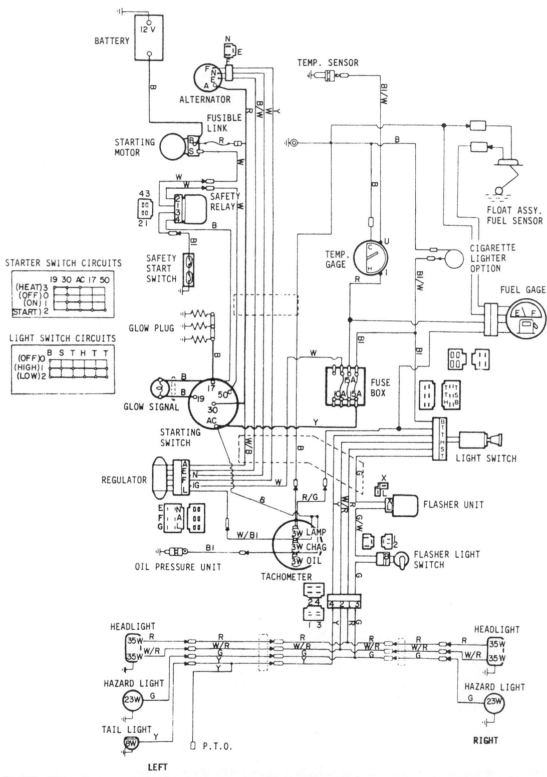

Fig. 310 — Wiring diagram typical of late production (July 1985 and after) 1910 models. Refer to Fig. 309 for wire color code.

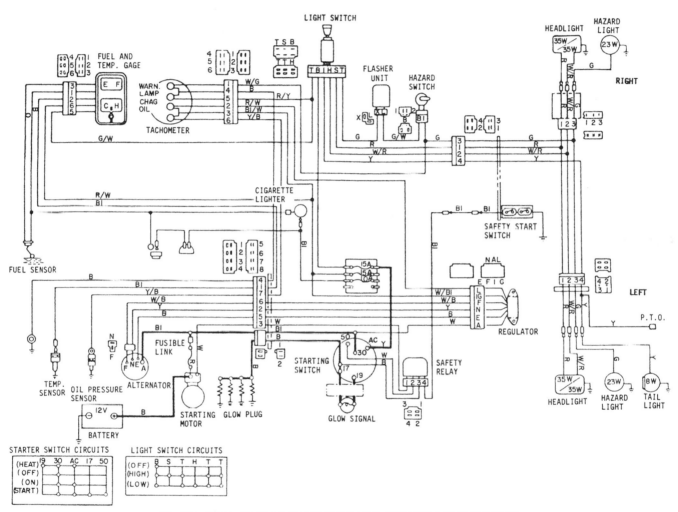

Fig. 311 — Wiring diagram typical of 2110 models. Wire color code is as follows:

B. Black	W. White	R/Y. Red/yellow stripe	W/G. White/green stripe
BL. Blue	Y. Yellow	W/B. White/black stripe	W/R. White/red stripe
G. Green	BL/W. Blue/white stripe	W/BL. White/blue stripe	Y/B. Yellow/black stripe
R. Red	R/W. Red/white stripe		

NOTES

NOTES

NOTES